Signal Analysis in Power Systems

Signal Analysis in Power Systems

Editor

Zbigniew Leonowicz

MDPI • Basel • Beijing • Wuhan • Barcelona • Belgrade • Manchester • Tokyo • Cluj • Tianjin

Editor
Zbigniew Leonowicz
Wroclaw University of Science and Technology
Poland

Editorial Office
MDPI
St. Alban-Anlage 66
4052 Basel, Switzerland

This is a reprint of articles from the Special Issue published online in the open access journal *Energies* (ISSN 1996-1073) (available at: https://www.mdpi.com/journal/energies/special_issues/Signal_Analysis).

For citation purposes, cite each article independently as indicated on the article page online and as indicated below:

LastName, A.A.; LastName, B.B.; LastName, C.C. Article Title. *Journal Name* **Year**, *Article Number*, Page Range.

ISBN 978-3-03936-820-4 (Hbk)
ISBN 978-3-03936-821-1 (PDF)

Contents

About the Editor

Zbigniew Leonowicz received his M.S. and Ph.D. degrees in electrical engineering from the Wroclaw University of Science and Technology, Wroclaw, Poland, in 1997 and 2001, respectively, and his Habilitation degree from the Bialystok University of Technology, in 2012. From 1997, he was with the Electrical Engineering Faculty, Wroclaw University of Technology. He also received two titles of Full Professor in 2019 from the President of Poland and the President of the Czech Republic. Since 2019, he has been a Professor at the Department of Electrical Engineering, where he is currently the Head of the Chair of electrical engineering fundamentals. He has received Best Reviewer Awards multiple times from international journals, including *Elsevier Electric Power System Research* and *Publons*.

Preface to "Signal Analysis in Power Systems"

This issue is devoted to reviews and applications of modern methods of signal processing used to analyze the operation of a power system and evaluate the performance of the system in all aspects. Monitoring capability with data integration, advanced analysis of support system control, and enhanced power security are the key issues discussed in the paper "Analysis of the Power Supply Restoration Time after Failures in Power Transmission Lines". Advanced statistical analysis of the power system is presented in papers "Combined Cluster Analysis and Global Power Quality Indices for the Qualitative Assessment of the Time-Varying Condition of Power Quality in an Electrical Power Network with Distributed Generation" and "The Application of Hierarchical Clustering to Power Quality Measurements in an Electrical Power Network with Distributed Generation", demonstrating the cutting-edge developments in this emerging area. The relatively new concept of virtual power plants, related to ongoing research in cooperation with industrial partners from the energy sector is presented in the paper "A Case Study on Distributed Energy Resources and Energy-Storage Systems in a Virtual Power Plant Concept: Technical Aspects". New concepts of photovoltaic energy forecasting complete the issue with the paper "Forecasting Solar PV Output Using Convolutional Neural Networks with a Sliding Window Algorithm".

Zbigniew Leonowicz
Editor

energies

MDPI

Article

Forecasting Solar PV Output Using Convolutional Neural Networks with a Sliding Window Algorithm

Vishnu Suresh *, Przemyslaw Janik, Jacek Rezmer and Zbigniew Leonowicz

Faculty of Electrical Engineering, Wroclaw University of Science and Technology, 50-370 Wroclaw, Poland; przemyslaw.janik@pwr.edu.pl (P.J.); jacek.rezmer@pwr.edu.pl (J.R.); zbigniew.leonowicz@pwr.edu.pl (Z.L.)

* Correspondence: vishnu.suresh@pwr.edu.pl

Received: 25 December 2019; Accepted: 5 February 2020; Published: 7 February 2020

Abstract: The stochastic nature of renewable energy sources, especially solar PV output, has created uncertainties for the power sector. It threatens the stability of the power system and results in an inability to match power consumption and production. This paper presents a Convolutional Neural Network (CNN) approach consisting of different architectures, such as the regular CNN, multi-headed CNN, and CNN-LSTM (CNN-Long Short-Term Memory), which utilizes a sliding window data-level approach and other data pre-processing techniques to make accurate forecasts. The output of the solar panels is linked to input parameters such as irradiation, module temperature, ambient temperature, and windspeed. The benchmarking and accuracy metrics are calculated for 1 h, 1 day, and 1 week for the CNN based methods which are then compared with the results from the autoregressive moving average and multiple linear regression models in order to demonstrate its efficacy in making short-term and medium-term forecasts.

Keywords: convolutional neural networks; multi-headed CNN; CNN-LSTM; forecasting; solar output; sliding window; renewable energy

---

1. Introduction

Global efforts to keep the increase in average temperature below 2 °C, with the possibility of keeping it lower than 1.5 °C, was agreed upon in the Paris agreement of 2015. In the recent "Climate action and support trends—2019" report, it was mentioned that current greenhouse gas emission levels and reduction efforts are not in line with meeting the targets that were set out [1].

Due to such environmental concerns and ambitious targets, there has been an increasing penetration of renewable energy sources in the power sector, especially in the form of solar photovoltaic panels. One of the biggest concerns connected with solar energy is its stochastic nature and variability, which threatens grid stability. A well-known approach to mitigate such uncertainty is the use of accurate forecasts [2].

The motivation for this study is the need to build a forecasting algorithm for a stochastic energy management system for the microgrid present at the Wroclaw University of Science and Technology. The microgrid currently employs a system that is deterministic, but, considering the stochastic nature of the solar panels, it was considered necessary. Convolutional neural network-based architectures used in forecasting are mainly used to study images of the sky, as explained later, and are used in tandem with statistical techniques for forecasting. This microgrid facility does not possess a device to record images of the sky but a deep learning approach to forecasting was decided upon. Hence, a data level approach using the sliding window algorithm for forecasting was adopted and the results were analyzed.

The area of forecasting is widely researched and is an age-old concept, aiming to predict solar PV outputs, wind turbine power outputs and loads in an electrical power system. A short literature review reveals numerous approaches, some of which are described as follows. In [3], short-term forecasts for PV outputs were obtained using Support Vector Regression models wherein the parameters of the models were optimized using intelligent methods, such as the Cuckoo Search and Differential Evolution algorithms. In this study, the authors had used data from an inhouse rooftop solar PV unit at Virginia Tech. In [4], multiple linear regression was employed to make forecasts for solar energy output. This study had used extensive data obtained from the European Centre for Medium-Range Weather forecasts, including as many as 12 independent variables. The study described in [5] presents a generalized fuzzy logic approach in order to make short-term output forecasts from measured irradiance data. The input data in this case was for one particular month (October 2014) and the inputs and outputs were normalized within a range of 0.1–0.9. A comprehensive review and analysis of different methods and associated results regarding the forecasting of solar irradiance and solar PV output is presented in [6].

With regard to the application of Convolutional Neural Networks (CNN) for solar PV output forecasts, there is little available literature. One of the approaches as seen in [7,8] is to use a combination of historical data and sky images. The sky images are crucial in order to capture the effect clouds have on PV output. The study described in [8] used a total sky imager, which provides images of the sky and could, whereas [7] used videos recorded by a 6-megapixel 360 degrees fish eye camera by HiKvision. Other approaches, which do not use images but only historical data, have adjusted the CNN in such a way that it is able to deal with time series data. CNN is, in fact, a machine learning tool that is explicitly used for image detection and classification but based on the method by which data is processed, its ability to understand non-linear relationships between the inputs and outputs can be leveraged for time series data. A hybridized approach where CNN is used for pattern recognition and then a long short-term memory network is used for prediction is seen in [9] and then this framework is applied for 30 min ahead forecasting of global solar radiation. In [2], a method in which suitable data processing is applied before training the CNN is presented. In this case, the time series data is split into various frequencies through variational mode decomposition and it is then converted into a 2D data form that is extracted by convolutional kernels. Finally, the approach used in [10] proposes another hybrid method in which a chaotic Genetic Algorithm/Particle Swarm Optimization is used to optimize the hyper parameters of the CNN, which is then used to make solar irradiance prediction.

This paper's forecasting approach is to be applied in developing a stochastic energy management system for microgrids. Hence, a few contributions in this regard are as follows: p A comprehensive review about weather forecasts, forecast errors, data sources, different methodologies used, and their importance in microgrid scheduling is described in [11]. The focus has been kept on wind energy forecasts, solar generation, and load forecasts. Another popular approach for forecasts using the ARMA (Autoregressive moving average) model, especially for load forecasting followed by solving a microgrid unit commitment problem, is described in [12]. An advanced forecasting method using artificial neural networks, support vector regression, and random forest followed by incorporation into a Horizon 2020 project involving several countries has been described in [13].

This paper utilizes a sliding window approach in order to prepare data in such a way that it can be used to train the CNN with historical data and make accurate predictions.

2. Forecasting Models, Data Processing, and Evaluation Metrics

2.1. Forecasting Models

The data for this study comes from a PV panel installed at a university building of the Wroclaw University of Science and Technology. It is a part of a power plant with a peak power capacity of 5 kW. The input measurements are obtained from associated sensors and are Irradiation (W/m^2), Wind speed (m/s), Ambient temperature (°C), and PV Module Temperature (°C). The output of the panel (W) and all inputs are measured in a 15 min window. The forecasting is also done in steps of 15 min intervals.

The inputs were chosen according to the recommendations of the IEA (International Energy Agency) report on "Photovoltaic and Solar Forecasting" [14] and other reliable sources [15]. The evaluation and benchmarking techniques to be used for the forecasts were also taken from [14–16] in order to establish the reliability of the results of this study. The metrics are discussed in detail further on.

The structure of the CNN model is shown in Figure 1.

The CNN is a specialized neural network that is explicitly used for image recognition. In such cases, the input images are represented as a two-dimensional grid of pixels. In order to use CNNs for time series data, a 1-D structure is more appropriate. Taking the example of the input time series data used in this study, it is a 175,200 × 4 matrix. The length (number of rows) represents the time step of the input data, whereas the columns (Irradiation, Wind Speed, Ambient temperature, PV Module Temperature) represent the width. This can be equated to the height and width of the pixels that are used as the input data for training CNNs for image recognition.

For efficient and quick training of all networks, the min–max scaling algorithm was used. This is necessary since the distribution and scale of the data varies for every variable. Moreover, the units of measurement for every variable are also different, which could lead to large weight values, and models assuming such large weight values often perform poorly while learning and are sensitive to changes in input values [17]. It was applied to normalize the data within the range of [0,1]. The formula for the same is described in (1).

$$\frac{x_i - \min(x)}{\max(x) - \min(x)} \tag{1}$$

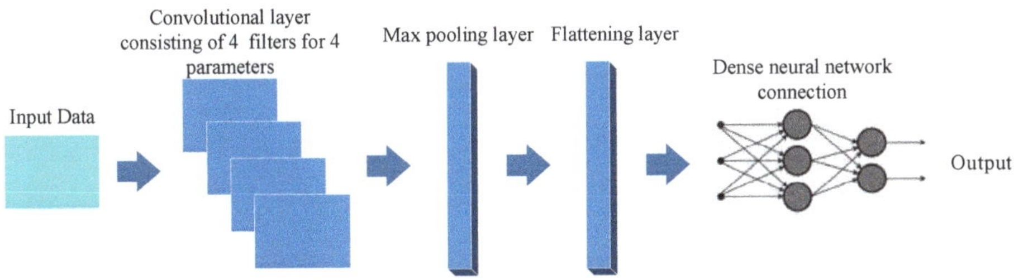

Figure 1. Convolutional Neural Network (CNN) structure.

The convolutional layer that follows input data processing is responsible for feature extraction [18]. The layer is made up of as many filters (neurons) as there are variables (4). These filters carry out convolution, which, by definition, is a function that is applied to the input data to obtain specific information from it. These filters are moved across the entire input data in a sliding window-like manner. In case of 2-D images the sliding window is moved horizontally and vertically but since this study employs a 1-D data the window is made to move vertically. The function used in this case is the Rectified Linear Activation Function (RLAF), which is described below, and the sliding window algorithm is described later.

The RLAF is a function that behaves like a linear function but is actually non-linear in nature, which enables the learning of complex relationships in the input data. It is widely used and can be defined in an easy manner. When the input is greater than 0.0, the output value remains the same as the input value, whereas if the input is less than 0.0 the output is 0.0. Mathematically, it is defined as:

$$g(z) = \max\{0, z\} \tag{2}$$

where z is the input value and g is the RLAF function. The advantage of this function includes computational ease, sparsity, and the ease of implementation to neural networks due to its linear behavior despite non-linearity [19].

The output of the filters in the convolutional layers are called feature maps. The feature maps hold relationships and patterns from the input data. These feature maps from each filter put together complete the convolutional layer. This layer is followed by the pooling layer, the objective of which is to reduce the feature maps of the convolutional layer (it summarizes the features learnt in the previous layer). This is done in order to prevent overfitting. It also reduces the size of the input data, which results in increased processing speeds and reduced memory demand. While there are numerous pooling functions, such as max, average, and sum [18], this study employs the max function, hence the max pooling layer.

The flattening layer succeeding the max pooling layer converts the output into a 1-D input vector that can be given to the dense or fully connected layer. The dense layer in this case is a regular neural network that has a non-linear activation function.

The model in this case is fit by the Adam optimization algorithm. The advantage of using this optimizer is that the learning rate is adjusted as the error is reduced.

It is in fact a combination of two well-known extensions of stochastic gradient descent, which are the Adaptive gradient algorithm (AdaGrad) and Root mean square propagation (RMSProp). Adam is discussed in detail in [20].

The second CNN structure used in this study is the multi-headed CNN. This approach involves handling every input series by its own CNN. This approach has shown some flexibility. While there is no significant proof in the literature behind the advantages of multi-headed CNN over the regular CNN using multiple filters, a multi-headed CNN with 3 convolutional 2-D nets has been used for enhanced image classification as shown in [21]. This paper uses a similar, yet different, architecture. The structure of the multi-headed CNN is shown in Figure 2.

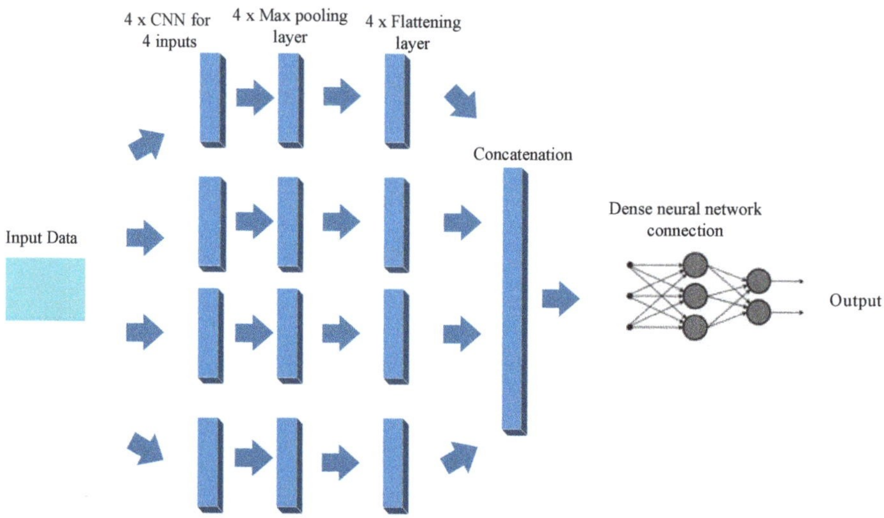

Figure 2. Multi-headed CNN structure.

In this study, as described in Figure 2, the multi-headed CNN has 4 CNNs, one for each input. This is followed by 4 max pooling layers and then by 4 flattening layers, and then the results from these layers is combined before the information is fed to the dense neural network, which makes the final prediction.

The third approach for forecasting is the CNN-LSTM (CNN-Long Short-Term Memory) network. Recently, the CNN-LSTM has been implemented in many areas for time series predictions. Study [22] presents a problem where water demand in urban cities is predicted. The correlation between water demand and changes in temperature and holiday periods is obtained using CNN-LTSM networks, and an improvement in predictions was observed. Similarly, an improvement in weather predictions was demonstrated in [23] by using such a hybrid CNN-LSTM architecture.

The LSTM is in fact an RNN (Recurrent neural network), which is efficient in working with time series data and is known to be a powerful tool for classification and forecasting associated with time series data. The uniqueness of LSTM comes from the memory cell, which behaves as a collector of state information. Whenever new information is obtained if the input gate is triggered it will be accumulated in the cell and past information would be forgotten if the forget gate is triggered. The latest cell obtained in such a process would be propagated to the final stage only if the output gate is triggered. This kind of cell behavior prevents the gradients trapped in the cell from vanishing quickly and is characteristic of LSTM, which makes it better suited to handle time series data and make predictions compared to other RNN structures [24].

The advantage of using a hybrid CNN-LSTM architecture is that the CNN is used to extract features from the raw input time series and then these features are given as an input to the LSTM, which is efficient with time series data.

Figure 3 provides the CNN-LSTM architecture. It can be noticed that, overall, the structure is similar to the CNN structure in Figure 1, with the exception of the LTSM layer, which enables the whole network to process the time series data more efficiently.

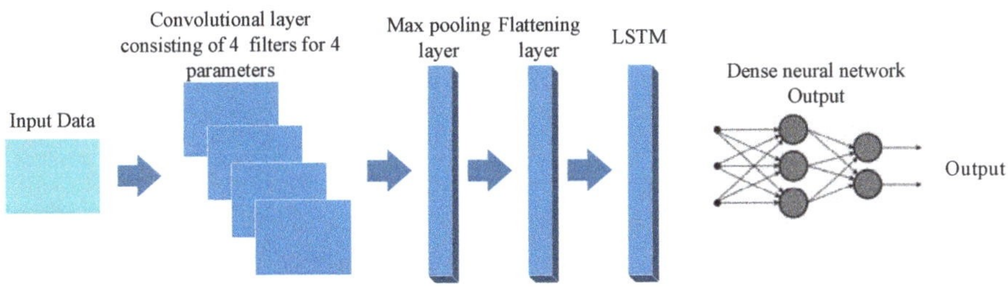

Figure 3. CNN-LSTM structure.

In order to provide a benchmark with an established technique for forecasting, the ARMA model is proposed. The ARMA model is utilized mainly for stationary time series data. In this method, the predicted variable is calculated on the basis of a linear relationship with its past values [25,26]. In cases when the data is non-stationary and has seasonal characteristics, as will be explained in the next section, it has to be transformed into a stationary one before an ARMA model is fit. The model consists of two parts, AR (Autoregressive) and MA (Moving Average), and is defined as ARMA (m, n) where m, n represent the orders of the model.

$$y_t^{'AR} = \sum_{i=1}^{m} \varnothing_i x_{t-i} + \omega_t = \varnothing_1 x_{t-1} + \varnothing_2 x_{t-2} + \ldots + \varnothing_m x_{t-m} + \omega_t \quad (3)$$

$$y_t^{'MA} = \sum_{j=0}^{n} \theta_j \omega_{t-j} = \omega_t + \theta_1 \omega_{t-1} + \theta_2 \omega_{t-2} + \ldots + \theta_n \omega_{t-n} \quad (4)$$

$$y_t^{'ARMA} = \sum_{i=1}^{m} \varnothing_i x_{t-i} + \sum_{j=0}^{n} \theta_j \omega_{t-j} \quad (5)$$

where $y_t^{'AR}$, $y_t^{'MA}$, and $y_t^{'ARMA}$ represent the time series values of the autoregression (AR), the Moving average (MA), and the Autoregression moving average (ARMA), respectively. $\varnothing_i$ is the autoregressive coefficient and θ_j is the moving average coefficient. ω_t is the noise.

The autoregressive (AR) part involves representing the current value as a result of a linear combination of the previous values and the noise ω_t. It is represented in Equation (3). The Moving average part is a combination of previous individual noise components, which is used to create a time series, as shown in Equation (4). ARMA is a combination of both AR and MA [27].

The parameters of the model m and n are chosen on the basis of an auto correlation function (ACF) and a partial auto correlation function (PACF). The ACF provides a correlation between a value of a given time series with past values of the same series, whereas the PACF provides a correlation between a value of the time series with another value at a different lag. If the ACF is reduced to a minimum value after a few lags and PACF depicts a large cut-off after the initial value, the time series is said to be stationary. This is then finally confirmed by the Augmented Dickey Fuller (ADF) test, which is explained in [25]. A confidence level of 95% is assumed for this study, hence a p-value of less than 0.05 is a confirmation of stationarity.

The analysis of the time series data according to the ACF, PACF, and ADF, in addition to its conversion to a stationary time series followed by the fitting of an ARMA model, is discussed in the next section.

Finally, the same data is also fit with a linear regression model. The linear regression model is explained below. A comprehensive study on the use of linear regression along with an improved model for hourly forecasting can be found in [28].

$$Y = \beta_o + \beta_1 X_1 + \beta_2 X_2 + \ldots + \beta_k X_k + \epsilon \tag{6}$$

where Y is the dependent variable, X_k are the independent variables, β_0 is the constant term, β_k is the coefficient corresponding to the slope of each independent variable, and ϵ is model's error, also known as residuals

2.2. Data Processing (Sliding Window)

While using the sliding window data processing approach for CNNs, a time series dataset is split as follows. The input data column is split into vectors consisting of an equal number of time steps. So, assuming the input data has 10 time steps, it is split into 5 vectors consisting of 2 time steps each. Then, these vectors are mapped to a label that is an output value from the training data. In this way, 5 vectors are mapped to 5 output values and 5 values are dropped, resulting in a reduced computational burden during the training of the model. The algorithm for the sliding window approach is presented in Algorithm 1.

Algorithm 1 sliding window

```
Procedure Variables (X, V, t)
i = 0, n = 0;                      # number of windows = n
K = [];                            # K is the set of windows extracted
While i + V ≤ length (X) do        #V is the length of the sliding window
K[n] = X [i . . . . (i + V − 1)};
i = i + t; n = n + 1;
end While
return F
end Procedure
```

While a general definition of the sliding window algorithm is presented here, every CNN model needs data to be prepared according to its structure. The sliding window for the CNN model in this study is applied to multivariate (the presence of more than one variable for every time step) time series data. In this case, every window determined by the algorithm has 2-time steps and its associated variables mapped to one output. The multi-headed CNN has 4 convolutional layers for every available input variable, hence the input time series is split into 4 univariate (one variable per time step) time series for each convolutional layer. Then, the sliding window algorithm is applied to each univariate series, and every window determined by the algorithm has 2-time steps and its associated variable mapped to an output.

The CNN-LSTM model reads input data in a different manner. In this case, the first step involves the application of the sliding window, where every window determined has 4-time steps, and then it is reshaped into 2 sub sequences containing associated variables and is mapped to outputs. The window is applied to a multivariate time series data.

2.3. Evaluation Metrics

The evaluation metrics chosen for this study were chosen based on recommendations of studies and reports in the field of solar PV output forecasting [6,14]. The metrics are the Root Mean Square Error (RMSE), Mean absolute error (MAE), and Mean Bias Error (MBE). RMSE is a metric that is widely used in forecast studies. According to [29], it is suitable for such data since it has the tendency to punish the largest errors with the largest effect, which the MAE and the MBE are unable to do. MAE is calculated as the average of the forecast errors. The MBE also calculates the average forecast errors but does not take in the absolute magnitude alone, this gives information regarding whether the model has a tendency to over or under forecast. The metrics are as follows:

$$RMSE = \sqrt{MSE} = \sqrt{\frac{1}{N}\sum_{i=1}^{N} e_i^2} \quad (7)$$

$$MAE = \frac{1}{N}\sum_{i=1}^{N} |e_i| \quad (8)$$

$$MBE = \frac{1}{N}\sum_{i=1}^{N} e_i \quad (9)$$

$$e_i = y_{i(forecast)} - y_{i(observed)} \quad (10)$$

where $y_{i(forecast)}$ and $y_{i(observed)}$ represent the forecasted and observed observations at the ith time step. e_i is the error at i^{th} time step. $i = 1, \ldots\ldots, N$ represents all the time steps within the data.

The evaluation metrics presented in the results section were calculated on the basis of original data after normalized prediction values were converted back using the inverse of the min–max scaling algorithm presented in Equation (1).

3. Results

All models were built on PYTHON using jupyter notebook. The deep learning tools that were used are TensorFlow and KERAS where the models were assembled. Additionally, Sci-kit learn and other basic Python libraries were used for data processing and data handling. The computer used for this purpose was equipped with an Intel®Core™ i5-4210 U CPU@ 1.70 GHz 2.40 GHz processor with an installed 8 GB of RAM operating Windows 10. It was also equipped with a 2048 MB GeForce 840M Nvidia graphics card. The training times for the CNN, Multi-CNN, and the CNN-LSTM models were 1364 s, 1657 s, and 3534 s, respectively. All architectures used the same data stretching over 6 years for model training and were trained for 100 epochs. The ARMA and MLR models were fit quite instantaneously, providing an advantage over the CNN based models with regard to the computational cost involved in model fitting. Once the models are fit, they are quite easy to use for the purposes of predictions. There is not any significant difference in terms of ease of usage amongst the statistical and CNN based techniques. Both models would need re fitting from time to time in order to take into account the changes in climate.

The data used for training the models were 6+ years' worth of data recorded from 1 March 2012 up to 31 December 2018. The validation split (test/train split) used was 20%, meaning that 80% of the data was used to train the CNN models and 20% was used to test them. The evaluation metrics obtained for 1 h, 1 day, and 1 week for both summer and winter months were obtained by testing the model for the months of July and December in 2019, which was unknown to the training models. There was no validation split for the MLR and ARMA models. They were fit on to the whole data and were tested with the July and December data of 2019, same as for the CNN models.

Figure 4 represents the time series data used in this study for the ARMA model without the validation split, it is quite evident that data has seasonality where the peaks in power output are observed during the summer. Hence, the periodicity for this study would be taken as 12 months. A look at the ACF with 20 lags indicates significant correlation. In fact, a clear pattern is visible when the lags are further increased to 60 and above. The PACF of the data also does not show any large cut-offs after the initial value hence the time series is non-stationary and has to be converted to a stationary time series before the ARMA model is fit to the data.

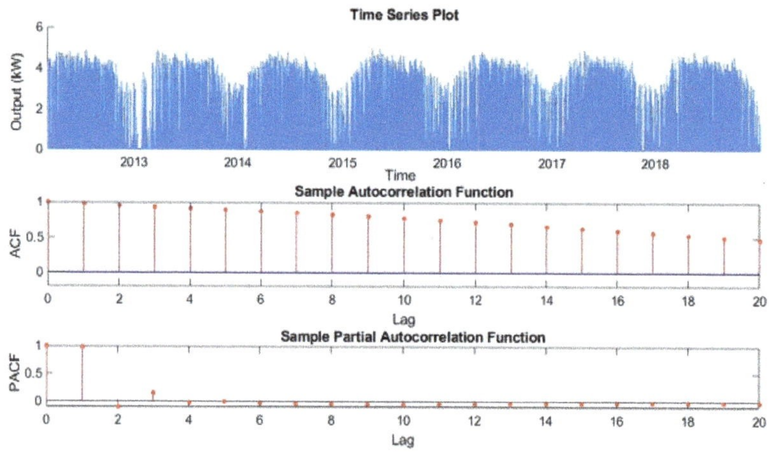

Figure 4. Solar panel output data with an auto correlation function (ACF) and a partial auto correlation function (PACF) analysis.

Figure 5 presents the differentiated time series. It can be seen from its characteristic that it fluctuates around zero, which is a defining characteristic for a stationary signal. Furthermore, in comparison with Figure 4, it can be seen that the ACF is not significant and also does not possess a trend, which is also the case for the PACF. In both cases, there is a sharp cutoff at 12, indicating seasonality at 12, which is in line with the selection of seasonality or periodicity at 12.

The ADF test made with the differentiated signal resulted in a p-value of 0.001, which confirms that the signal is stationary. Now the ARMA model parameters can be determined since the ACF and PACF are negligible beyond lag 2 therefore m and n could have a maximum value of 2. In this study, the m and n are taken as 1 and 2 and the following ARMA model is obtained.

Table 1 presents the ARMA model parameters that are used to predict solar output values for an hour, 1 day, and 1 week. The model ignores the constant value due to its high p-value. The evaluation metrics for the model predictions are presented in Table 2, and a comparison of the predictions as a result of model application with other methods is shown later on. Figure 6 presents the manner in which an appropriate forecasting model is obtained by different CNN architectures used. Figure 6a represents the loss value that is optimized in every epoch for the multi-headed CNN structure. It can be observed that for this model there is not any improvement in reduction of the loss function over many epochs of training. After an initial drop in the loss value it remains a constant, which means that training the architecture for a small number of epochs is sufficient for an accurate model. Figure 6b represents the loss value minimization for a simple CNN structure. In contrast to the multi-headed CNN structure, the loss minimization is more gradual, yet in a small number of epochs, a satisfying model is obtained. It has been noticed during several trials that, in the simple CNN structure, the loss minimization keeps improving up to a 1000 epochs and more. However, the improvement in forecast accuracy is not significant vis-à-vis the time it takes to train the model for a high number of epochs.

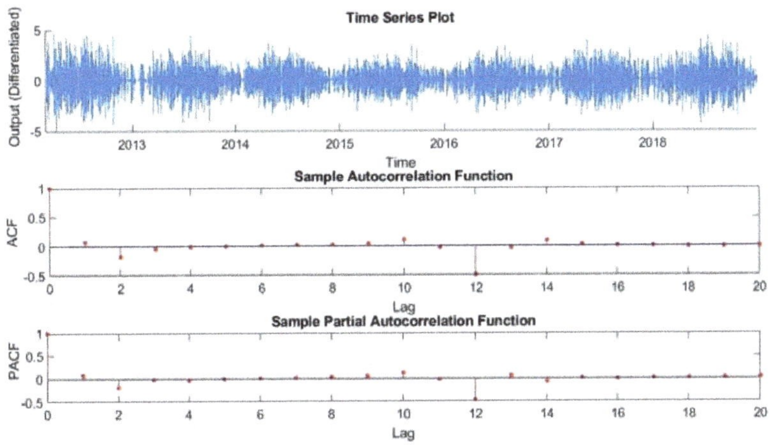

Figure 5. Differentiated output with ACF and PACF analysis.

Table 1. Autoregressive moving average (ARMA) model Parameters.

Parameter	Estimated Value	p-Value
$\varnothing_1$	0.208	0.000
θ_1	−0.125	0.000
θ_2	−0.197	0.000
Constant term	0.000	1.000
Variance	0.107	0.000

$\varnothing_1$—AR coefficient 1, θ_1—MA coefficient 1, θ_2—MA coefficient 2.

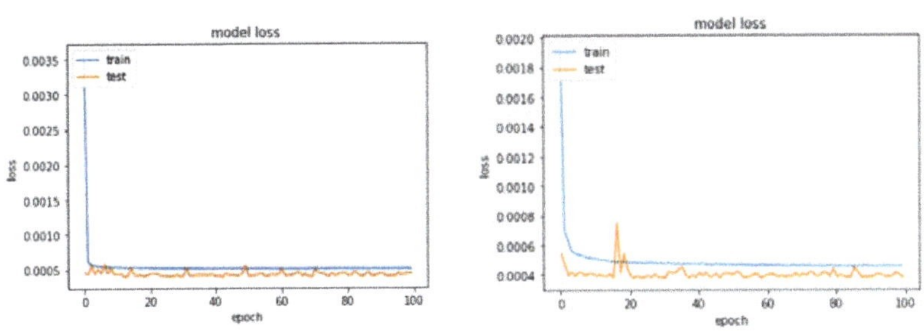

(**a**) (**b**)

Figure 6. Model fitting test and train loss minimization for (**a**) Multi-headed CNN (**b**) and Simple CNN structure.

Figure 7 represents the loss value minimization for the CNN-LSTM architecture. In comparison with Figure 6a,b, it can be observed that the model fitting takes slightly longer, yet it is completed with sufficient accuracy within 20 epochs. The model keeps improving with an increasing number of epochs, but it has been observed that, with a higher number of epochs (>500), the model tends to overfit with the loss curves of the test and train the data crossing over one another. For comparison purposes, keeping in mind the time for model fitting, 100 epochs was considered to be sufficient for all models.

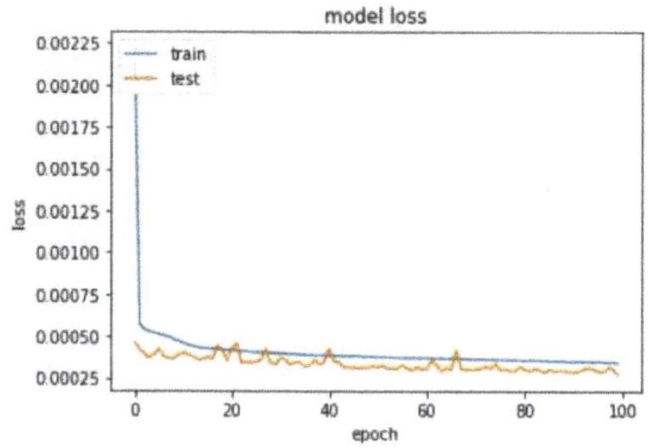

Figure 7. Model fitting test and train loss minimization for the CNN-LSTM network.

Table 2 presents the various metrics, as described in the previous section, which help understand the accuracy of the forecasts. The metrics are calculated for one hour (h), 1 day (D), and 1 week (W) in order to understand its consistency over the short and medium term. Table 2 is specifically for the summer months and the week in question is the 1st week of July 2019.

Table 2. Forecast metrics for the short and long term during the summer months.

Methods Used	RMSE (1 h)	MAE (1 h)	BIAS (1 h)	RMSE (1 D)	MAE (1 D)	BIAS (1 D)	RMSE (1 W)	MAE (1 W)	BIAS (1 W)
CNN-Simple	0.068	0.066	−0.066	0.051	0.033	−0.017	0.056	0.031	−0.016
Multi-headed CNN	0.169	0.169	−0.169	0.081	0.053	−0.036	0.080	0.050	−0.038
CNN-LSTM	0.053	0.053	−0.053	0.051	0.035	−0.025	0.045	0.030	−0.019
ARMA	0.046	0.043	+0.043	0.192	0.153	+0.153	0.244	0.134	+0.880
Multiple linear regression	0.477	0.474	−0.474	0.258	0.179	−0.149	0.258	0.146	−0.120

The RMSE, MAE, BIAS values are all in kW, hour—h, day—D, week—W.

It can be noticed that the values of RMSE, MAE, and the BIAS for the 1h forecast for all CNN-based methods are nearly the same. This is because the number of observations within an hour is just limited to four (due to 15 min time step) and the neural network methods take multiple inputs in order to make one prediction. In fact, for the CNN-LSTM model, the inputs needed are four for one prediction, hence the RMS, MAE, and the BIAS are the same for the 1 h forecast. It can be noticed from the BIAS value that, in the case of all methods being used except the ARMA model, there is a slight tendency to overpredict. The ARMA model has performed as good as any used CNN method and has the most accurate prediction followed by the CNN-LSTM.

For the 1-day forecasting (2 July 2019), it can be seen that the CNN-simple and the CNN-LSTM perform in a similar manner. The RMSE being around 0.051 kW. All methods have shown better performance than the MLR. The BIAS still indicates a tendency to over predict except for the ARMA model. It can be noticed that while the ARMA model provided very accurate results for the 1-h predictions, the RMSE value has increased considerably for the 1-day forecasts. The multi-headed CNN performs the worst amongst the CNN models.

For the 1-week forecasting (1st week of July) it can be noticed that the CNN-LSTM makes more accurate forecasts than the CNN-simple and multi-headed CNN models. In general, it can be noticed that with longer forecasts the accuracy metrics improve, indicating a general improvement or at least consistency in predictions for the CNN-based models. On the contrary for the ARMA model, the RMSE value has increased considerably from the 1-h and 1-day predictions. The BIAS in this case is to overpredict, except for the ARMA model.

Figure 8 presents the 1-day forecasting (2 July 2019) made by different algorithms explained previously. It can be inferred from the Figure that the most accurate forecasts for the day are made by the CNN and the CNN-LSTM models, wherein they almost overlap the actual values. These are followed by the multi-headed CNN model, which has a higher error in its predicted values, the ARMA model, which, despite being more accurate than MLR, underpredicts at moments of sharp changes.

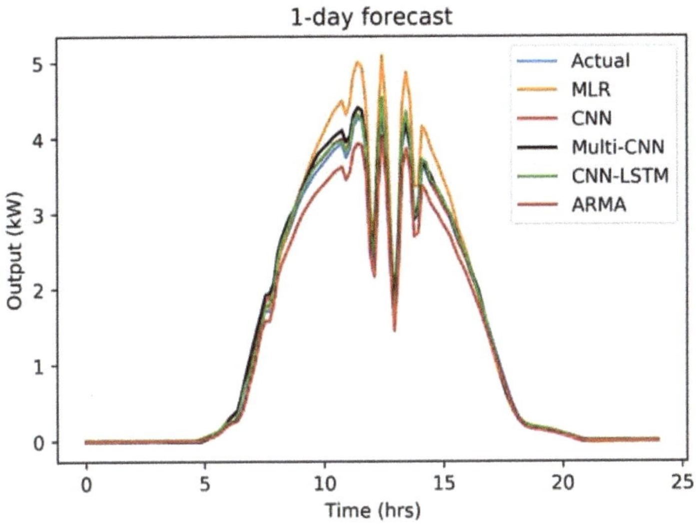

Figure 8. Comparison of different methods for the 1-day forecast (summer).

Figure 9 represents the forecasts made by different approaches used for a week (the 1st week of July). It can be noticed that for most parts the forecasts form CNN-LSTM, CNN-simple (CNN), and Multi-CNN closely match the actual values, though, at the peak, some inaccuracy can be noticed with the multi-CNN forecasts. The ARMA model slightly underpredicts, especially during peak output. The MLR is also inaccurate at the peaks.

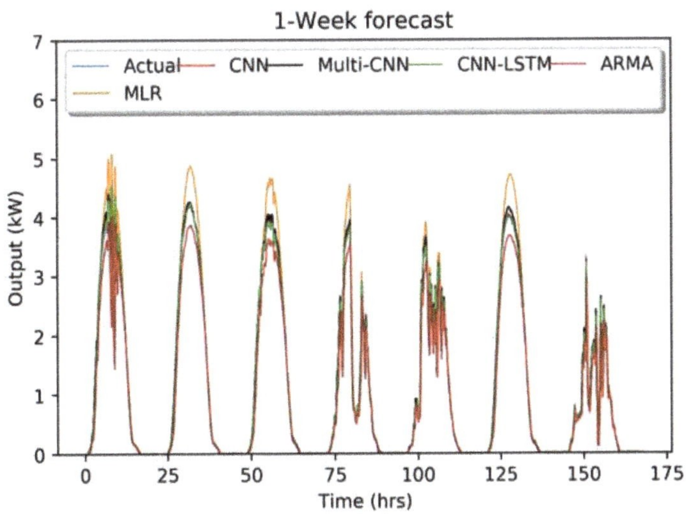

Figure 9. Comparison of different methods for the 1-week forecast (summer).

Table 3 provides the evaluation metrics of the models used during the winter months, it can be seen that the models have less accuracy for the 1-h predictions when compared to their own predictions during the summer. Again, the ARMA model performs with the highest accuracy for the 1-h predictions followed by the CNN-LSTM model. For the longer 1-day (28 December 2019) and 1-week periods (3rd week of December) it can be seen that the performance of all CNN-based models is very consistent with their performance during the summer months. The most accurate forecasting is made by the simple CNN model for 1-day (28 December 2019) forecasting, whereas for the 1-week forecasting it is the CNN-LSTM model. An intriguing observation between the summer and winter models is the fact that the difference between the RMSE and MAE values is higher during the winter period. The RMSE values are two times that of the MAE values and are higher for the CNN-based methods for the 1-day predictions, they are almost 3 times higher during the 1-week predictions, which is an indication that when errors are made during predictions they are higher in magnitude when compared to predictions in the summer because the RMSE has the inherent characteristic to give more weight to bigger errors.

Table 3. Forecast metrics for the short and long term during the winter months.

Methods Used	RMSE (1 h)	MAE (1 h)	BIAS (1 h)	RMSE (1 D)	MAE (1 D)	BIAS (1 D)	RMSE (1 W)	MAE (1 W)	BIAS (1 W)
CNN-Simple	0.466	0.451	+0.450	0.036	0.018	+0.005	0.104	0.039	+0.026
Multi-headed CNN	0.341	0.328	+0.328	0.038	0.019	−0.016	0.092	0.032	+0.023
CNN-LSTM	0.297	0.295	−0.295	0.056	0.029	+0.005	0.100	0.036	+0.022
ARMA	0.188	0.187	+0.187	0.040	0.018	+0.018	0.142	0.050	+0.023
Multiple linear regression	0.465	0.778	+0.778	0.092	0.059	−0.011	0.040	0.098	+0.032

The RMSE, MAE, and BIAS values are all in kW, hour—h, day—D, week—W.

Figure 10 presents the forecasting made for 1 day (28 December 2019) during the winter. In comparison with the 1-day prediction for the summer, the ARMA and MLR forecast values have significantly improved, and all methods predict quite accurately.

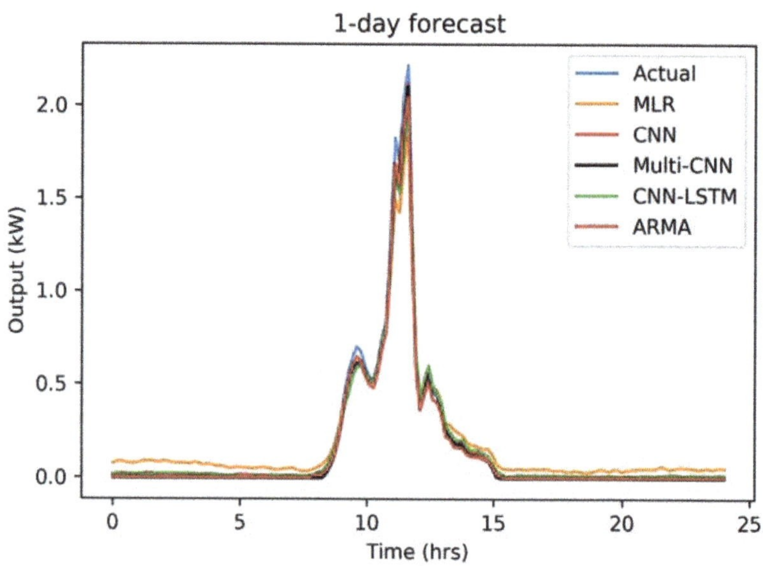

Figure 10. Comparison of different methods for the 1-day forecast (winter).

Figure 11 represents the forecasting made by different approaches used in this study for the 3rd week of December, it can immediately be noticed that the output values are lower when compared with the summer week, with one day having almost no output. Except for the MLR, it can be seen that all models predict quite closely to the real values, with all of them underpredicting a little during peak power output.

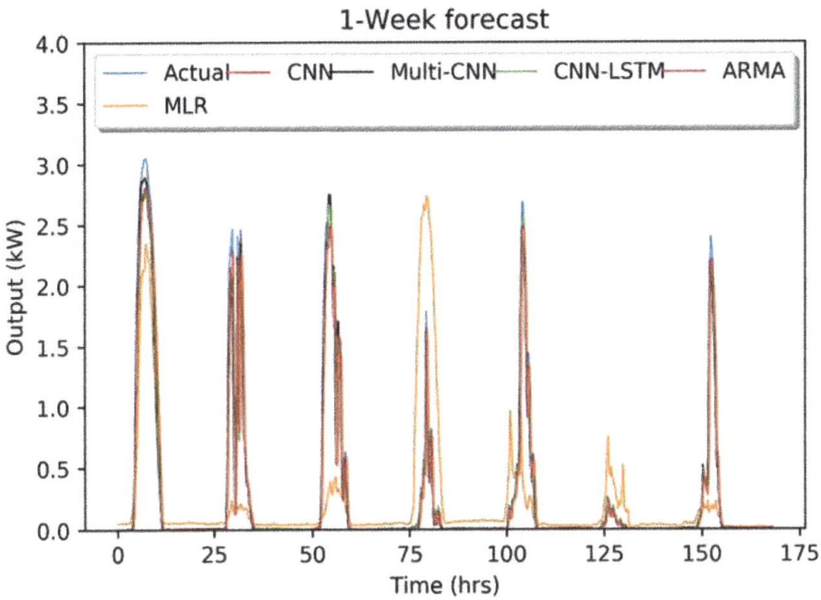

Figure 11. Comparison of different methods for the 1-Week forecast (winter).

4. Discussion

This paper presents a forecasting approach using deep learning neural networks. The neural network structures, used primarily for image recognition, have been adapted to handle time series data with a seasonal characteristic. In order to make this possible, a data processing approach, such as the sliding window algorithm, has been used. A comparison between the performance of different possible structures of the neural network has been carried out along with a multiple linear regression and ARMA model. It has been noticed that the CNN-simple and the CNN-LSTM methods perform best for all 1-h, 1-day and 1-week predictions, with the CNN-LSTM providing better results on certain occasions. The ARMA model performed exceptionally for the 1-h forecasts. The forecasting was carried out for 1 h, 1 day, and 1 week with the function of electricity markets in mind. From the accuracy metrics such as RMSE, MAE, and BIAS it can be concluded that the forecasting algorithms perform satisfactorily. Since its performance has been tested in the short and medium term at the university location, it will be followed by rigorous testing at other locations in order to establish its applicability across geographical regions with different seasonal characteristics. Future work in this regard includes investigating the performance of other architectures that possess more abstraction (the level of complexity of the neural network). Abstraction can be increased by increasing the number of convoluted layers, which may or may not improve the accuracy metrics. It also has an effect on the training time for fitting an appropriate model. Furthermore, a different combination of CNNs and RNNs could be considered. Additionally, the effect of clustering is to be explored. The CNN-based models were fit, and, on the whole, the model performs quite uniformly across all seasons, which we believe is due to the fact that during the training method a set of values from the past is used to train the model at every step. This enables the model to capture seasonality as the weather-related variables from past values are clearly season-dependent. The approach is similar in the ARMA model since it also uses past values, but it is important to remember that the ARMA class of models are applicable only to univariate data (in this case the output values of the PV plant). Moreover, this study is a part of building a stochastic Energy Management System for microgrids, hence they will be used as inputs for optimization algorithms managing the EMS.

Author Contributions: Conceptualization and methodology were developed by authors V.S. and P.J., software, validation, formal analysis, investigation was done by V.S., resources, data curation and writing—original draft preparation was done by V.S. and J.R., writing—review and editing, visualization, supervision, project administration was done by author P.J. and Z.L. All authors have read and agreed to the published version of the manuscript.

Funding: This research received funding from the Chair of Electrical Engineering Fundamentals (K38W05D02), Wroclaw University of Technology, Wroclaw, Poland.

Conflicts of Interest: The authors declare no conflict of interest.

References

1. UNCCS. Climate action and support trends, United Nations. *Clim. Chang. Secr.* **2019**, 7.
2. Zang, H.; Cheng, L.; Ding, T.; Cheung, K.W.; Liang, Z.; Wei, Z.; Sun, G. Hybrid method for short-term photovoltaic power forecasting based on deep convolutional neural network. *IET Gener. Transm. Distrib.* **2018**, *12*, 4557–4567. [CrossRef]
3. Alrashidi, M.; Alrashidi, M.; Pipattanasomporn, M.; Rahman, S. Short-Term PV Output Forecasts with Support Vector Regression Optimized by Cuckoo Search and Differential Evolution Algorithms. In Proceedings of the 2018 IEEE International Smart Cities Conference, Kansas City, MO, USA, 16–19 September 2019; pp. 1–8. [CrossRef]
4. Abuella, M.; Chowdhury, B. Solar power probabilistic forecasting by using multiple linear regression analysis. In Proceedings of the IEEE Southeastcon, Fort Lauderdale, FL, USA, 9–12 April 2015; pp. 1–5. [CrossRef]
5. Chugh, A.; Chaudhary, P.; Rizwan, M. Fuzzy logic approach for short term solar energy forecasting. In Proceedings of the 2015 Annual IEEE India Conference (INDICON), New Delhi, India, 17–20 December 2015; pp. 1–6. [CrossRef]
6. Yang, D.; Kleissl, J.; Gueymard, C.A.; Pedro, H.T.C.; Coimbra, C.F.M. History and trends in solar irradiance and PV power forecasting: A preliminary assessment and review using text mining. *Sol. Energy* **2018**, *168*, 60–101. [CrossRef]
7. Sun, Y.; Venugopal, V.; Brandt, A.R. Short-term solar power forecast with deep learning: Exploring optimal input and output configuration. *Sol. Energy* **2019**, *188*, 730–741. [CrossRef]
8. Ryu, A.; Ito, M.; Ishii, H.; Hayashi, Y. Preliminary Analysis of Short-term Solar Irradiance Forecasting by using Total-sky Imager and Convolutional Neural Network. In Proceedings of the 2019 IEEE PES GTD Grand International Conference and Exposition Asia (GTD Asia), Bangkok, Thailand, 20–23 March 2019; pp. 627–631. [CrossRef]
9. Ghimire, S.; Deo, R.C.; Raj, N.; Mi, J. Deep solar radiation forecasting with convolutional neural network and long short-term memory network algorithms. *Appl. Energy* **2019**, *253*, 113541. [CrossRef]
10. Dong, N.; Chang, J.-F.; Wu, A.-G.; Gao, Z.-K. A novel convolutional neural network framework based solar irradiance prediction method. *Int. J. Electr. Power Energy Syst.* **2020**, *114*, 105411. [CrossRef]
11. Agüera-Pérez, A.; Palomares-Salas, J.C.; de la Rosa, J.J.G.; Florencias-Oliveros, O. Weather forecasts for microgrid energy management: Review, discussion and recommendations. *Appl. Energy* **2018**, *228*, 265–278. [CrossRef]
12. Alvarado-Barrios, L.; del Nozal, Á.R.; Valerino, J.B.; Vera, I.G.; Martínez-Ramos, J.L. Stochastic unit commitment in microgrids: Influence of the load forecasting error and the availability of energy storage. *Renew. Energy* **2020**, *146*, 2060–2069. [CrossRef]
13. Alamo, D.H.; Medina, R.N.; Ruano, S.D.; García, S.S.; Moustris, K.P.; Kavadias, K.K.; Zafirakis, D.; Tzanes, G.; Zafeiraki, E.; Kaldellis, J.K.; et al. An Advanced Forecasting System for the Optimum Energy Management of Island Microgrids. *Energy Procedia.* **2019**, *159*, 111–116. [CrossRef]

14. Luukkonen, P.; Bateman, P.; Hiscock, J.; Poissant, Y.; Dignard-Bailey, L. International Energy Agency Co-Operative Programme on Photovoltaic Power Systems. 2013. Available online: http://www.cansia.ca/sites/default/files/201306_cansia_2012_pvps_country_report_long.pdf (accessed on 29 October 2019).
15. Paulescu, M.; Paulescu, E.; Gravila, P.; Badescu, V. Weather Modeling and Forecasting of PV Systems Operation. *Green Energy Technol.* **2013**, 325–345. [CrossRef]
16. Lorenz, E.; Remund, J.; Müller, S.C.; Traunmüller, W.; Steinmaurer, G.; Pozo, D.; Ruiz-Arias, J.A.; Fanego, V.L.; Ramirez, L.; Romeo, M.G.; et al. Benchmarking of Different Approaches to Forecast Solar Irradiance. *Eur. Photovolt. Sol. Energy Conf* **2009**, 4199–4208. [CrossRef]
17. Bishop, C.M. *Pattern Recognition and Machine Learning*; Springer: Cambridge, UK, 2006. [CrossRef]
18. Bisong, E. Convolutional Neural Networks (CNN). In *Building Machine Learning Deep Learning Models Google on Cloud Platform*; Apress: Berkeley, CA, USA, 2019; pp. 423–441. [CrossRef]
19. Heaton, J. Ian Goodfellow, Yoshua Bengio, and Aaron Courville: Deep learning. *Genet. Program. Evolvable Mach.* **2018**, *2018 19*, 305–308. [CrossRef]
20. Kingma, D.P.; Ba, J. Adam: A Method for Stochastic Optimization. *arXiv* arXiv:1412.6980, 2014.
21. Mathur, P.; Ayyar, M.P.; Shah, R.R.; Sharma, S.G. Exploring classification of histological disease biomarkers from renal biopsy images. In Proceedings of the 2019 IEEE Winter Conference on Applications of Computer Vision (WACV), Waikoloa Village, HI, USA, 7–11 January 2019; pp. 81–90. [CrossRef]
22. Hu, P.; Tong, J.; Wang, J.; Yang, Y.; de Oliveira Turci, L. A hybrid model based on CNN and Bi-LSTM for urban water demand prediction. In Proceedings of the 2019 IEEE Congress on Evolutionary Computation, Wellington, New Zealand, 10–13 June 2019; pp. 1088–1094. [CrossRef]
23. Fu, Q.; Niu, D.; Zang, Z.; Huang, J.; Diao, L. Multi-stations ' Weather Prediction Based on Hybrid Model Using 1D CNN and Bi-LSTM. In Proceedings of the 2019 Chinese Control Conference (CCC), Guangzhou, China, 27–30 July 2019; pp. 3771–3775.
24. Shi, X.; Chen, Z.; Wang, H.; Yeung, D.Y.; Wong, W.K.; Woo, W.C. Convolutional LSTM network: A machine learning approach for precipitation nowcasting. In *Advances in Neural Information Processing Systems*; Cornell: New York, NY, USA, 2015; pp. 802–810.
25. Atique, S.; Noureen, S.; Roy, V.; Subburaj, V.; Bayne, S.; MacFie, J. Forecasting of total daily solar energy generation using ARIMA: A case study. In Proceedings of the 2019 IEEE 9th annual computing and communication workshop and conference (CCWC), Las Vegas, NV, USA, 7–9 January 2019; pp. 114–119. [CrossRef]
26. Alsharif, M.H.; Younes, M.K.; Kim, J. Time series ARIMA model for prediction of daily and monthly average global solar radiation: The case study of Seoul, South Korea. *Symmetry* **2019**, *11*, 240. [CrossRef]
27. Ghofrani, M.; Suherli, A. Time Series and Renewable Energy Forecasting. *Intech* **2016**, *13*, 77–92. [CrossRef]
28. Trigo-González, M.; Batlles, F.J.; Alonso-Montesinos, J.; Ferrada, P.; del Sagrado, J.; Martínez-Durbán, M.; Cortés, M.; Portillo, C.; Marzo, A. Hourly PV production estimation by means of an exportable multiple linear regression model. *Renew. Energy* **2019**, *135*, 303–312. [CrossRef]
29. Madsen, H.; Pinson, P.; Kariniotakis, G.; Nielsen, H.A.; Nielsen, T.S. Standardizing the performance evaluation of short-term wind power prediction models. *Wind Eng.* **2005**, *29*, 475–489. [CrossRef]

energies

MDPI

Article

Combined Cluster Analysis and Global Power Quality Indices for the Qualitative Assessment of the Time-Varying Condition of Power Quality in an Electrical Power Network with Distributed Generation

Michał Jasiński [1,*], Tomasz Sikorski [1], Paweł Kostyła [1], Zbigniew Leonowicz [1] and Klaudiusz Borkowski [2]

1 Department of Electrical Engineering Fundamentals, Faculty of Electrical Engineering, Wroclaw University of Science and Technology, 50-370 Wroclaw, Poland; tomasz.sikorski@pwr.edu.pl (T.S.); pawel.kostyla@pwr.edu.pl (P.K.); zbigniew.leonowicz@pwr.edu.pl (Z.L.)

2 KGHM Polska Miedź S.A, 59-301 Lubin, Poland; klaudiusz.borkowski@kghm.com

* Correspondence: michal.jasinski@pwr.edu.pl; Tel.: +48-713-202-022

Received: 16 March 2020; Accepted: 13 April 2020; Published: 20 April 2020

Abstract: This paper presents the idea of a combined analysis of long-term power quality data using cluster analysis (CA) and global power quality indices (GPQIs). The aim of the proposed method is to obtain a solution for the automatic identification and assessment of different power quality condition levels that may be caused by different working conditions of an observed electrical power network (EPN). CA is used for identifying the period when the power quality data represents a different level. GPQIs are proposed to calculate a simplified assessment of the power quality condition of the data collected using CA. Two proposed global power quality indices have been introduced for this purpose, one for 10-min aggregated data and the other for events—the aggregated data index (*ADI*) and the flagged data index (*FDI*), respectively. In order to investigate the advantages and disadvantages of the proposed method, several investigations were performed, using real measurements in an electrical power network with distributed generation (DG) supplying the copper mining industry. The investigations assessed the proposed method, examining whether it could identify the impact of DG and other network working conditions on power quality level conditions. The obtained results indicate that the proposed method is a suitable tool for quick comparison between data collected in the identified clusters. Additionally, the proposed method is implemented for the data collected from many measurement points belonging to the observed area of an EPN in a simultaneous and synchronous way. Thus, the proposed method can also be considered for power quality assessment and is an alternative approach to the classic multiparameter analysis of power quality data addressed to particular measurement points.

Keywords: data mining; cluster analysis; power quality; global power quality index; electrical power network; distributed generation; mining industry

1. Introduction

Over the years, global electric energy consumption has increased from 440 Mtoe in 1973 to 1737 Mtoe in 2015 [1]. This has resulted in electricity becoming a specific product that is subject to market regulation in both quantitative and qualitative terms. Quantitative analysis is mainly focused on the balance between energy that is produced, transmitted, stored, consumed or lost. The current issues connected to quantitative aspects of energy consumption are related to demand-side response, the integration of renewable energy sources, and energy storage systems with electrical power systems [2–5]. The qualitative approach mainly uses power quality analysis. The issues related to power quality (PQ) include definitions of the parameters, and the methods of measurements and assessment, which are already standardized, among others [6–8]. The methods described in the standards of power quality analysis are based on power quality parameters measured during a representative period of time, normally one week, which corresponds to the normal working conditions of the observed network [9]. The parameters which characterize power quality include: frequency variation, voltage variation, voltage fluctuation, voltage asymmetry, and voltage waveform distortion. These parameters are collected during the period of observation, with the aggregation time interval usually equal to 10 min; however, 1 min aggregation intervals are also currently studied [10]. Using these parameters creates a significant number of data to be considered in the analysis. Moreover, PQ data depends on the network conditions, load changes, generation level, or configurations of the network. For this reason, a rational approach is to search for data mining techniques able to extract and classify vectors of the power quality data that represent different features. This would allow the range of qualitative analyses to be extended by correlating the information of the network, environment or market condition.

There are many works dedicated to power quality disturbance extraction, PQ events recognition, and the classification of PQ events and disturbances, which are all directly focused on the measured voltage and current signals. Most of the works propose wavelet transforms, S-transforms, empirical mode decomposition, and other different decomposition techniques, which are supported by artificial neural networks in order to find valuable methods for PQ disturbance extraction and recognition [11–13]. However, a different problem can be formulated when there is a need for identifying and extracting some of the data that represent different features from the long-term aggregated power quality data. This necessitates a comprehensive method for the automatic classification of long-term power quality data into groups that represent similar features. This task is essentially a data mining area of interest. One of the data mining techniques that can meet these requirements is cluster analysis (clustering). The general application of data mining techniques in power systems is presented in [14]. Specific applications of cluster analysis to electrical power networks include:

- Detection and classification of disturbances [15–19];
- Detection and analysis of events [20–25];
- Load characteristics and classification [26–30];
- Power flow problems [31–34];
- Power quality evaluation and monitoring [35–38].

This article extends the cluster analysis (CA) proposed by the authors in [9]. Jasiński et al. [9] present the results of the application of CA in order to achieve a desirable division of the long-term 10-min aggregated power quality data into groups of data representing similar features. The collection of the PQ data comes from four real points of measurement in the supply network of a copper mine. The significant elements of the investigated power network are combined heat and power (CHP) plants with gas-steam turbines working as a local distributed generation (DG), and also a welding machine (WM) as the main time-varying load. Time-varying PQ conditions were intentionally created. The distributed generation was switched on and off for a period of time, and a network reconfiguration was also performed. The results discussed in [9] confirm the possibility of using cluster analysis for the extraction of power quality data into groups related to the different working conditions of an electrical network, including the influence of DG, reconfiguration of the network, working days, and holiday time. In [9], the methodology of application of the cluster analysis, including the preparation of the database structure, was also described. The idea presented in [9] leads to efficient classification of the power quality data, but it does not provide a suitable method for the assessment of collected clusters of the data. Searching for (1) a comprehensive solution that provides automatic classification of the multipoint measurement data, and (2) a method for comparative evaluation of the collected data, remains a desirable aim for wide-area monitoring systems and smart grids. Thus this article is an extension of the previously obtained classification [9] in order to a achieve quality assessment of obtained clusters using global power quality indices. This leads to an automatic classification of working conditions of an electrical power network (EPN), and the possibility of an easy comparison using global values, that incorporate the impact of different PQ elements.

Both cluster analysis of PQ data and global power quality index (GPQI) application may be found in the literature:

- Sacasqui et al. [39] present an application of grey clustering with entropy weight methodology. The proposed solution was used to calculate a unified quality index of distributed electricity. Their research is based on [40], where a new unified index was proposed, as well as a network model. The model consists of a 138 kV system, wind energy system, hybrid wind-photovoltaic-fuel cell system and the load. The PQ data consist of current total harmonic distortion, voltage total harmonic distortion, sag, frequency deviation, instantaneous flicker level, and power factor. The unified index is calculated for different working conditions using gray CA and entropy weight for the measurement points separately. The research is based on simulations.
- The work of Song et al. [41] concerns the application of cluster analysis combined with a support vector machine for the prediction of PQ indexes. The real measurement data from a 35 kV substation are processed. The database contains selected PQ parameters including frequency deviation, voltage unbalance, and total harmonic distortion (*THD)* in voltage, as well as weather conditions and data on other associated factors. In the described article, CA was used to obtain implicit classifications of indexes. The analysis concerns a single measurement point.
- Florencias-Oliveros et al. [42] present the analysis of recorded signals representing different disturbances. The proposed index realizes a comparison of the variance values, skewness, and kurtosis connected with each cycle, versus the ideal signal. Then, the CA is used to create a classification of the disturbances using proposed PQ index.

The aspect that distinguishes the solution proposed in this paper from the methods described in quoted works is the area-based approach to the PQ assessment, involving all measurement points for the cluster analysis, as well as development of a new synthetic power quality index. Novel aspects of the method proposed in this article include:

- Application of cluster analysis for the data collected from several measurement points distributed in the supply network of a mining industry in order to achieve suitable identification of different working conditions of the observed network. This approach treats the collected data as a common database more representative of the observed area than particular measurement points.
- New synthetic global power quality indices are used for the assessment of groups of PQ data identified by cluster analysis. The proposed definition of the GPQI consists of a set of classical PQ parameters based on a 10-min aggregation interval; however, it is also extended by selected parameters based on a 200-ms aggregation interval. The aim of extending the proposed GPQI definition with parameters related to a 200-ms aggregation interval is to enhance the sensitivity of the obtained global index. This proposed approach is tested by investigating the influence of the factors which comprise the proposed global power quality index on the sensitivity of the assessment.
- The proposed approach of using GPQIs leads to a straightforward comparison of the clusters in terms of a generalized assessment of the power quality conditions, which in turn finally allows a comparative assessment of different working conditions of the investigated network to be performed. The indicated clusters, which represent different working conditions, may be easily compared using a single GPQI for each of the measurement points.

The remaining structure of this paper is as follows: Section 2 reviews the present application of global power quality indices in the electrical power network, and also proposes a new definition of the GPQIs proposed in our assessment of clustered PQ data. Section 3 describes the proposed algorithm methodology for the comparative assessment of the power quality conditions using a combination of clustering and global power quality indices. The first step of the algorithm is the identification and allocation of the power quality data into groups that represent similar features. This part is based on previous experience with CA application described in [9]. The second step is the assessment of the collected data using the proposed GPQI. The results of the assessment are presented using real multipoint power quality measurements in a medium voltage electrical network supplying the mining industry. Additionally, this section also contains a sensitivity analysis of the proposed GPQI in terms of the selection of the power quality parameters used to construct the GPQI. The presented results are towered to realize one of the article's aims—to highlight the impact of DG on PQ in the industry network. The obtained clusters represent different conditions of PQ indices which are directly associated with impact of the DG. Qualitative assessment of the PQ data collected in the identified clusters using the proposed global power quality indices allows us to confirm several relations between DG impact on PQ condition. Section 4 contains the discussion of the obtained results. Section 5 formulates the conclusions, interpretations in perspective studies, and implications for the future.

2. Global Power Quality Indices

Classical power quality assessment is a multi-criteria analysis approach that is independently applied to particular power quality parameters. The idea of a simplified and generalized assessment of the power quality condition uses a single index, known as a global, unified, total or synthetic index. In this paper, we decided to use global power quality indices (GPQIs) as a unified name. Before new definitions of GPQIs are introduced, it is relevant to have a review of the knowledge concerning the development of GPQIs. Singh et al. [43] present the application of a unified power quality index that uses the matrix method. The index, corresponding to voltage sag severity, was highlighted as a suitable proposition for power quality assessment, and is carried out in a three-stage approach. The first stage requires the preparation of a graphical system model (attribute digraph). The second step is the conversion into an attribute matrix. The next step is the presentation of the matrix as a variable permanent function. Ignatova and Villard [44] define green-yellow-red indicators for all PQ problems. The proposed algorithm obtains the green-yellow-red indicators for both events and disturbances. The index consists of all individual PQ parameters, which are expressed as a percentage in a range from 0% to 100%, where 0% denotes the worst PQ and 100% the optimal PQ. The index may be defined for each single point or for the whole facility. The benefit of the proposed generalization is the possibility to easily understand the interpretation of the PQ condition in the monitoring systems. Nourollah and Moallem [45] present the application of data mining to determine the unified power quality index which corresponds to all power quality parameters, with further classification, normalization, and incorporation. The proposed fast independent component analysis algorithm was proposed to determine the power quality level of each distribution site. The mentioned article proposes two indexes: the Supply-side Power Performance Index, which expresses the impact of six voltage indices; and the Load-side Power Performance Index. The second index corresponds to three current PQ indices. Raptis et al. [46] present artificial neural networks as a sufficient tool to support PQ assessment using an index called Total Power Quality Index. The index is the artificial neural network combination of eight power quality values used as input variables. The presented method uses a multilayer perceptron artificial neural network. Lee et al. [47] propose another power quality index. This index includes the power distortion, which concerns non-linear loads. The indicated aim of the proposed PQI is to support harmonic pollution determination in a distributed power system. The work [47] proposes a new distortion power quality index. The application of this index is a determination of the harmonic pollution ranking for different non-linear loads. It is realized by multiplication of the load composition rate and the load currents' total harmonic distortion. Hanzelka et al. [48] propose the idea of a synthetic PQ index. This index is based on the maximum values of traditional PQ parameters. These parameters are slow voltage change, harmonic content in voltage (represented by total harmonic distortion in voltage, and a particular harmonic from 2nd to 40th), unbalance, and voltage fluctuation (represented by long-term flicker severity). The proposed assessment provided only satisfactory or unsatisfactory results.

In the present work, two definitions of GPQIs are proposed—one for 10-min aggregated data, and the other for the events. The proposed indices are inspired by the synthetic approach described in [48,49]. Some elements of the GPQI definitions, in terms of the multipoint measurements, were also proposed by the authors in [50]. Typical for the generalization process is that global indices are usually less sensitive due to synthetization. In order to enhance the sensitivity, the global indices proposed in this work are not only based on classical 10-min aggregated power quality parameters, but they are also extended by other parameters like an envelope of voltage changes based on 200-ms values. In order to demonstrate the proposed approach, we also present an analysis of how selected parameters comprising the global index influence its sensitivity.

The first proposed global power quality index is called the aggregated data index (*ADI*), and is expressed in (1).

$$ADI = \sum_{i=1}^{7} k_i \cdot W_i \tag{1}$$

ADI—aggregated data index;

i—number of the factor ranging from 1 to 7;

W_i—the particular power quality factors which create a synthetic aggregated data index;

k_i—the importance rate (weighted factors) of the particular power quality factor constituting the synthetic aggregated data index, range of [0, 1], where $\sum_{i=1}^{7} k_i = 1$.

The *ADI* utilizes five classical 10-min aggregated PQ parameters, including: frequency (*f*), voltage (*U*), short-term flicker severity (P_{st}), asymmetry factor (k_{u2}), total harmonic distortion in voltage (*THDu*), and also two additional parameters which are responsible for the enhancement of the sensitivity of the proposed global index. The first additional parameter is represented by an envelope of voltage deviation obtained by the difference between the maximum and minimum of 200-ms voltage values identified during the 10-min aggregation interval. The second is a maximum of the 200-ms value of the total harmonic distortion in voltage, similarly identified in the 10-min aggregation interval. The mentioned parameters are calculated and refer to standard IEC 61000-4-30 [7]. Three phase values, like *U*, P_{st}, and *THDu* are reduced to one using the mean value of the three phase values. To be more specific, particular factors that create the proposed *ADI* index are based on the differences between the measured 10-min aggregated power quality data and the recommended limits stated in the standards. The differences are expressed as a percentage in relation to the limits. The final values of the factors taken in the *ADI* calculation are the mean values of the time-varying factors during the time period of observation. Additionally, the contribution of the particular power quality factors in global indices can be controlled by the importance factors, which serve as the weight of the contribution of particular parameters. The values of weighting factors are normalized to one. Selection of importance factors makes it possible to check the impact of single parameters as well as groups of parameters. The selection of parameters may be defined by a priori analysis of EPN problems (e.g., harmonics, voltage variations). No a priori statements were conducted in this work, so the weight of all parameters is the same and the priorities of particular parameters were the same. The aim of the introduced weighted factors is to open the possibility to make the analysis more focused on particular PQ parameters and neglect others—in other words, to obtain an analysis that is more sensitive for selected PQ phenomena controlled by weighted factors. For example, to justify adding 200-ms values, analyses with and without them were conducted.

Particular factors which create the global *ADI* index are defined as follows [50]:

$$W_1 = W_f = \frac{mean(|f_{\mathrm{m}} - f_{\mathrm{nom}}|)}{\Delta f_{\mathrm{limit}}} \tag{2}$$

$W_1 = W_f$—factor of frequency change;

f_{m}—10-min measured value of frequency;
f_{nom}—nominal value of frequency;
$mean\left(\left|f_{\mathrm{m}} - f_{\mathrm{nom}}\right|\right)$—mean of frequency deviations in the observation time period;
$\Delta f_{\mathrm{limit}}$—limit value of frequency change as a %.

$$W_2 = W_U = \frac{mean(|U_{\mathrm{m}} - U_{\mathrm{c}}|)}{\Delta U_{\mathrm{limit}}} \tag{3}$$

$W_2 = W_U$—factor of the voltage level;
U_{m}—mean of 10-min measured values of voltage from three phases;
U_{c}—declared voltage;
$mean(|U_{\mathrm{m}} - U_{\mathrm{c}}|)$—mean of voltage deviations in the observation period of time;
$\Delta U_{\mathrm{limit}}$—limit value of voltage change in volts.

$$W_3 = W_{Pst} = \frac{mean(Pst_{\mathrm{m}})}{Pst_{\mathrm{limit}}} \tag{4}$$

$W_3 = W_{Pst}$—factor of voltage variation;
Pst_{m}—mean of 10-min measurement value of the short-term flicker severity index from three phases;
$mean(Pst_{\mathrm{m}})$—mean of voltage variations in the observation time period;
Pst_{limit}—limit value of short-term flicker severity.

$$W_4 = W_{ku2} = \frac{mean(ku2_{\mathrm{m}})}{ku2_{\mathrm{limit}}} \tag{5}$$

$W_4 = W_{ku2}$—factor of voltage unbalance;
$ku2_{\mathrm{m}}$—10-min measured values of voltage unbalance;
$mean(ku2_{\mathrm{m}})$—mean value of voltage unbalance in the observation time period;
$ku2_{\mathrm{limit}}$—limit level of voltage unbalance.

$$W_5 = W_{THDu} = \frac{mean(THDu_{\mathrm{m}})}{THDu_{\mathrm{limit}}} \tag{6}$$

$W_5 = W_{THDu}$—factor of total harmonic distortion factor of voltage supply;
$THDu_{\mathrm{m}}$—mean of 10-min measurement values of the total harmonic distortion factor of the voltage supply from three phases;
$mean(THDu_{\mathrm{m}})$—mean value of the total harmonic distortion factor in the observation time period;
$THDu_{\mathrm{limit}}$—limit level of the total harmonic distortion factor of the voltage supply.

$$W_6 = W_{Uenv} = \frac{\frac{mean(|U_{\mathrm{max}} - U_{\mathrm{min}}|)}{U_{\mathrm{c}}}}{2 \times \Delta U_{\mathrm{limit}}} \tag{7}$$

$W_6 = W_{Uenv}$—factor of voltage deviation envelope;
U_{max}—mean value of 200-ms voltage maximum values from three phases allocated in 10-min data;
U_{min}—mean value of 200-ms voltage minimum values from three phases allocated in 10-min data;
U_{c}—declared voltage;
$mean(|U_{\mathrm{max}} - U_{\mathrm{min}}|)$—mean of voltage envelope width in the observation time period;
$\Delta U_{\mathrm{limit}}$—limit level of voltage change.

$$W_7 = W_{THDumax} = \frac{mean(THDu_{\mathrm{max}})}{THDu_{\mathrm{limit}}} \tag{8}$$

$W_7 = W_{THDu\max}$—factor of the maximum 200 ms value of the total harmonic distortion factor of voltage supply;

$THDu_{\max}$—mean value of 200-ms maximum values of the total harmonic distortion factor of voltage supply from three phases;

$mean(THDu_{\max})$—mean of the total harmonic distortion factor in the observation time period;

$THDu_{\text{limit}}$—limit level of the total harmonic distortion factor of the voltage supply.

Then, the preparation of the particular factors $W_1 \div W_7$ and the selection of its important rates, the aggregated data index factor expresses the PQ level in a global range. The interpretation of the obtained index values are as natural. A value of "0" represents the ideal PQ; "0–1" represents possible power quality deterioration, but in compliance with the requirements defined in the standards; and finally, a value greater than 1 indicates the permissible parameters level defined in the standard is exceeded.

The second proposed global index relates to events. The classical approach to power quality assessment utilizes a flagging concept, which generally prescribes the extraction of the aggregated values that are affected by events like dips, swells and interruptions. The authors propose to use the information about the number of data which are not considered in classical PQ analysis due to the flagging concept. This is used as the base for a global index called the flagged data index (*FDI*), defined as follows [50]:

$$FDI = \frac{f}{n} \times 100\% \tag{9}$$

FDI—flagged data index;

f—number of 10-min data, which were flagged in the observation time period;

n—number of all 10-min data in the observation time period.

Interpretation of the obtained *FDI* values can be formulated such that "0%" represents the ideal PQ without any event disturbances, and "100%" expresses measurement data where each averaged value is contaminated by voltage events.

The proposed concepts for the generalization of the power quality assessment using GPQIs can be implemented for the fixed time period of observation or for identified periods of time representing different features of the power quality condition of the monitored area of the power system. The identification of such periods can be achieved using cluster analysis.

3. Results of Power Quality Assessment Using Cluster Analysis and Global Power Quality Indices

The idea of combined analysis using CA and GPQIs is presented in Figure 1. In the first step, the clustering is applied to achieve a classification of the power quality data into clusters representing different features. The outcomes of the CA depend on the construction of the PQ database, that is the set of PQ parameters under consideration, as well as the standardization of the formula. The mentioned issues and their impact on the results of the CA were already investigated and presented in [9]. A novelty of this work is the implementation of GPQIs for the group of PQ data identified by CA. We propose using the levels of GPQIs that characterize particular clusters for the comparative analysis.

As was already mentioned, some results of the cluster analysis were described in [9]. However, selected information about the investigated electrical power network is repeated for clarity and to help in understanding the presented application of the global power quality indices. Note that the input PQ data that create the database are the four-week multipoint power quality measurements obtained from a 6 kV power network supplying the mining industry [51]. The points of measurement include a secondary side of 110 kV/6 kV transformers (denoted as "T1", "T2", "T3"), and a 6 kV outcoming feeder supplying a welding machine (denoted as "WM") [9]. Inside the network, distributed generation units are installed (denoted as "DG"), represented by combined heat and power plants (CHP) with gas-steam turbines, denoted as "G1", "G2", and "G3", respectively. The analyzed EPN of the mining industry and placement of the measurement points are presented in Figure 2.

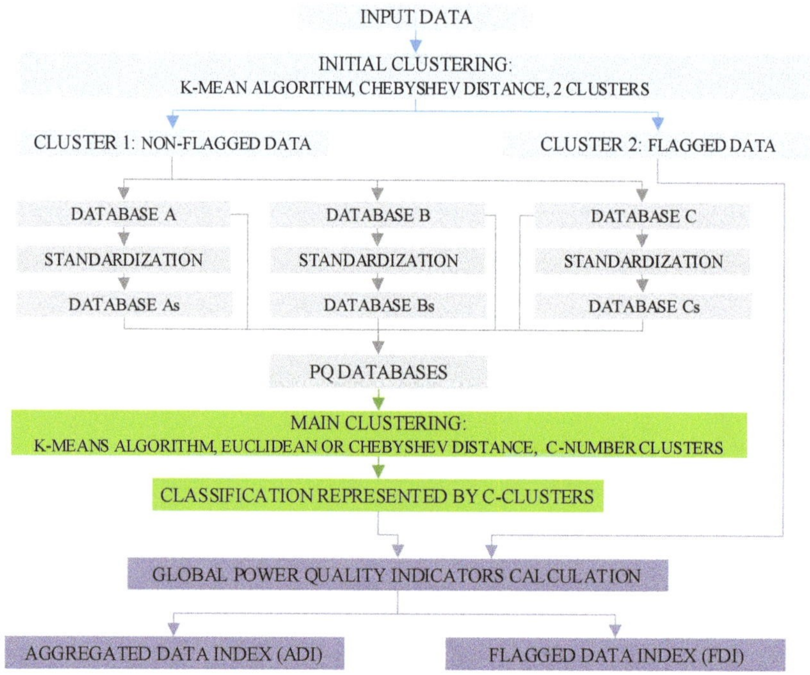

Figure 1. General scheme of the algorithm that performs the cluster analysis and global power quality (PQ) calculation.

The proposed method was implemented for the real measurements collected from four measurement points: three transformers T1, T2, T3 which supplied the medium voltage (MV) industrial network and a significant load (i.e., the welding machine—WM). The changes in the power demand of the investigated measurements points T1, T2, T3, and WM during the selected four weeks of observation are presented in Figure 3a. The investigation was aimed to evaluate the influence of the DGs installed inside the observed industrial network, and so Figure 3b presents changes in active power generation of particular DG units denoted as G1, G2, and G3. Generator G1 was permanently switched off during the experiment. G2 and G3 switched off, as can be seen in Figure 3b, due to a planned maintenance break. Additionally, it can be seen that during the experiment, only G2 (connected to the transformer T3 which also supplies the welding machine WM) and G3 (connected to transformer T2) were operating. The power variations of the DG is additional information, representing conditions. The data from the DG do not form the database of measurements taken for the investigation. An analysis of voltage events in the PQ measurements was conducted. Indicated events were voltage dips, rapid voltage changes, swells, and interruptions. Detailed information about the events and number of flagged data is included in Table 1 [52]. In accordance with the flagging concept introduced in the standard [7], the aggregated 10-min data that contained such voltage events were excluded from the power quality analysis. Based on the research presented in [9], it was shown that the best results of the CA with regards to the identification of different PQ conditions caused by the impact of the DGs could be achieved for the PQ databases denoted as C and C_S, where database C is constructed of frequency variation (f), voltage variation (U), short-term flicker severity (P_{st}), asymmetry (k_{u2}), total harmonic distortion in voltage ($THDu$), and active power level (P). Database C_S is the standardized version of database C, obtained by dividing the particular time series by their maximum values to achieve expression of the data in the range 0–1. Thus, for the investigation presented in this paper, database C and its standardized version Cs were taken for consideration.

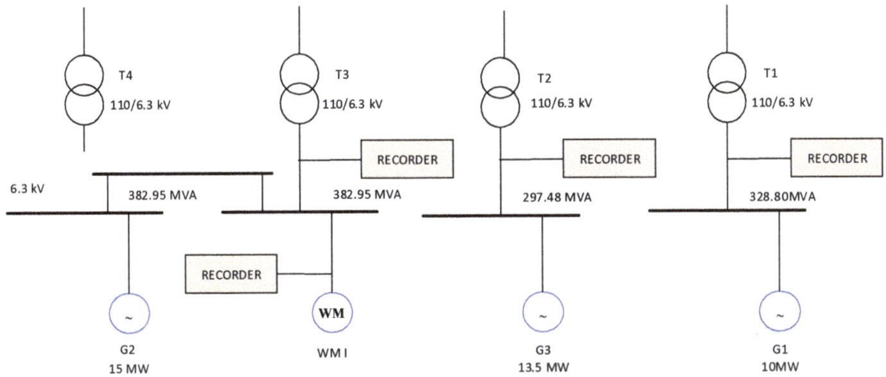

Figure 2. Analyzed mining industry electrical power network (EPN) and placement of distributed generation and PQ recorders [9]. T1—transformer 1; T2—transformer 2; T3—transformer 3; T4—transformer 4; G1—generator 1; G2—generator 2; G3—generator 3; WM—welding machine.

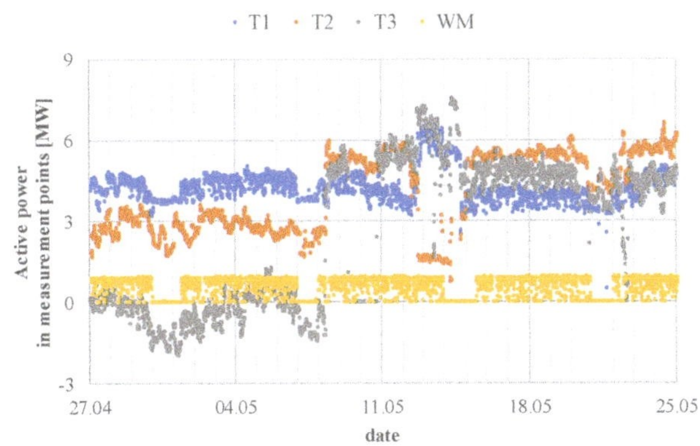

(a)

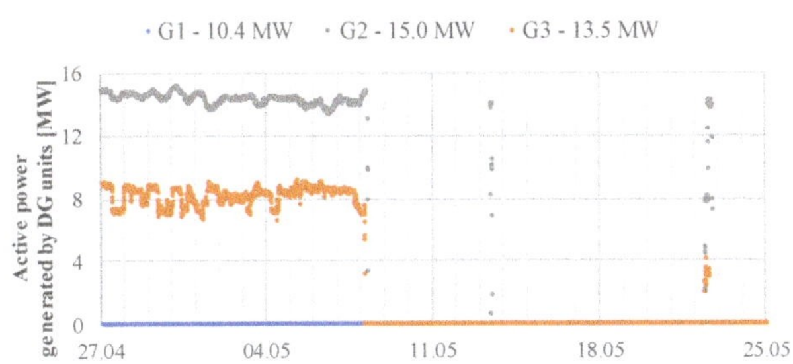

(b)

Figure 3. The analysis of active power level in electrical power network of mining industry: (**a**) Active power of the observed point of the measurements in the investigated network including the high voltage/medium voltage (HV/MV) transformers T1, T2, T3 and the connection point of the welding machine WM; (**b**) Active power of the distributed generators (DGs) during the investigated period of observation (G2 is connected to transformer T3, G3 is connected to transformer T2).

Table 1. Indication of events and number of flagged data during the measurements.

Point	Type of Event	Flag Start		Flag Stop		Number of 10-min Flagged Data
		Date	Hour	Date	Hour	
T1	long interruption, voltage dip, swell, rapid voltage change	21.05	07:50:00	21.05	15:10:00	45
T2	voltage dip	05.05	05:40:00	05.05	05:40:00	1
	long interruption, short interruption, voltage dip, swell, rapid voltage change	14.05	08:20:00	14.05	16:20:00	49
	voltage dip	05.06	14:00:00	05.06	14:00:00	1
T3	voltage dip	05.05	05:30:00	05.05	05:30:00	1
	rapid voltage change	07.05	22:40:00	07.05	22:40:00	1
	long interruption, voltage dip, swell, rapid voltage change	09.05	09:00:00	10.05	17:20:00	195
	long interruption, voltage dip, transient overvoltage, rapid voltage change	20.05	08:00:00	20.05	19:20:00	69
WM	voltage dip	05.05	05:30:00	05.05	05:30:00	1
	rapid voltage change	07.05	22:40:00	07.05	22:40:00	1
	short interruption, voltage dip, rapid voltage change	20.05	07:40:00	20.05	07:40:00	1

3.1. Cluster Analysis—Identification of the Power Quality Data Representing Different PQ Conditions Due to the Impact of DG

In [9], different results of the clustering were presented using different numbers of clusters (2, 3, 5, 20). It was shown that increasing the number of clusters enabled the identification of data not only related to the impact of the DG (i.e., when the DG was active or switched off), but also for the extraction of data associated with other working conditions (i.e., working day or non-working day, time of the network reconfiguration). This article aims to highlight the influence of distributed generation on power quality in the industry network. Thus, referring to the achievements presented in [9], in this work the scope of the CA was limited to the aim of classifying the data into three groups: cluster 1—DG was active; cluster 2—DG was switched off; cluster 3—other conditions. After the experiences described in [9], we decided to use the K-means algorithm with Euclidean distance.

In order to visualize the association of the obtained clusters with the distributed generation work information, Figure 4, which presents the clustering results, is supported by additional, artificial clusters indicated as cluster −1 and cluster 0, which were created on the basis of external information collected by the control and monitoring systems of particular DGs, as well as the output of the PQ monitoring systems considering the flagged data. Cluster −1 denotes the time series when the DG was active. This approach enables the easy comparison of the CA outcomes with regards to the identification of the working condition of the DGs. As was previously indicated, the databases are comprised only of unflagged data. Cluster 0 concerns flagged data that must be excluded from the main cluster analysis. The main clusters that are the outcomes of the CA analysis are cluster 1, which represents data when the DG was working, and cluster 2, which expresses the time period when the DG was switched off. Comparing the outcomes of the applied clustering with an artificial informative cluster denoted as −1 allows for the conclusion that the applied technique provides an appropriate output for connection of the clusters to different working states of the DG time period. Figure 4 presents the outcomes of the clustering with Euclidean distance when the initial number of clusters is 3. Referring to the information coming from external network dispatcher systems, it was confirmed that the time period indicated by cluster 3 was related to the reconfiguration of the network topology. In this case, increasing the number of clusters ensures the determination of a more sensitive classification of the collected PQ data when a specific working condition of the EPN is indicated. These and other issues concerning the initial number of clusters and the construction of the database were studied in [9]. However, it is important to note that the clustering is the first step in the multipoint long-term measurement analysis, which ensures a classification of the data into groups that are matched with the specific condition of the observed network. It finally leads to the possibility of the qualitative assessment of the data collected into clusters, as well as comparative analysis between the clusters. For this purpose, this paper proposes the use of global power quality indices.

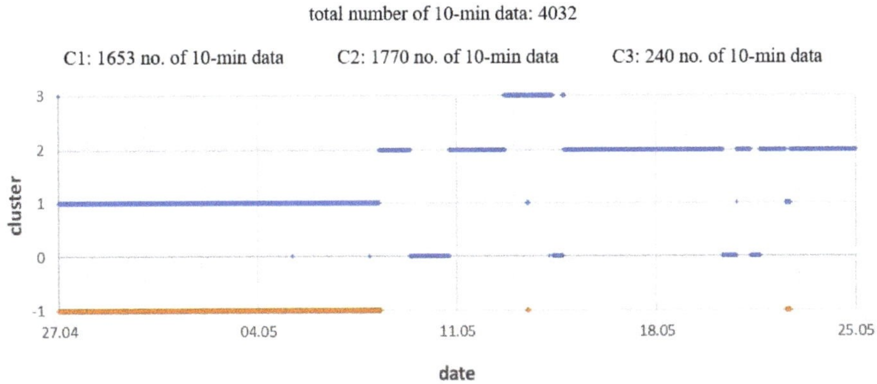

Figure 4. Results of power quality data clustering using cluster analysis (CA) with K-means and Euclidean distance and three initial clusters. C1—the distributed generation (DG) was working; C2—the DG was switched off; C3—DG was switched off and with a different network topology configuration.

3.2. *Qualitative Assessment of the Determined Clusters Based on the Proposed Global Power Quality Indices*

As was described in Section 2, the proposed aggregated data index (*ADI*) uses five components based on 10-min aggregated data, and two other components based on 200-ms data. The acceptance levels for the *ADI* components, with regards to aggregated power quality parameters, are presented in Table 2. The values correspond to the demands included in the standard [6].

Table 2. The acceptance level of the components of the *ADI* related to 10-min aggregated power quality parameters in reference to [6].

Parameter	Value
Δf_{limit}	0.5 Hz
ΔU_{limit}	10%
Pst_{limit}	1.2
$ku2_{\text{limit}}$	2%
$THDu_{\text{limit}}$	8%

In the presented results, each importance rate $k_1 \div k_7$ (weighted factors) of the seven parameters comprising the *ADI* were the same and equal to 1/7. This means that the importance of all the parameters was treated equally. The 10-min step *ADI* variation for particular measurement points (T1, T2, T3, WM) in relation to the determined clusters of the PQ data is presented in Figure 5. In order to link the *ADI* variation with the output of the CA analysis, that is time periods which refer to particle clusters, colored backgrounds for particular clusters were inserted in Figure 5.

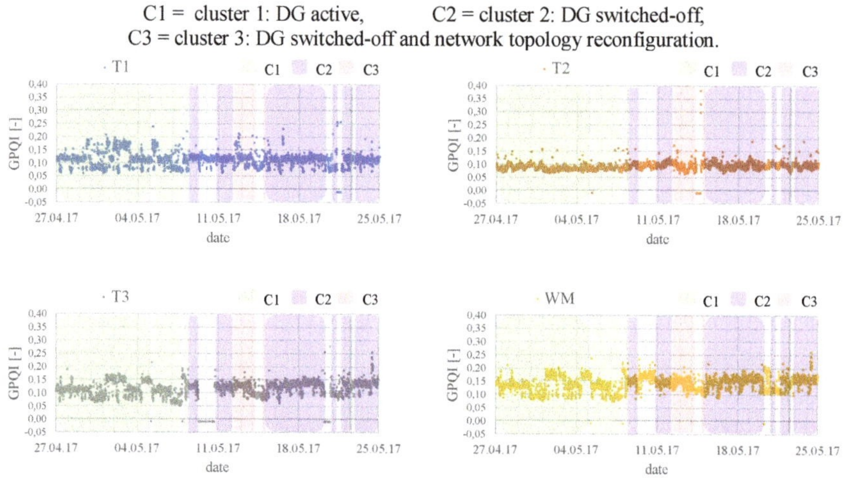

Figure 5. *ADI* variation for particular measurement points (T1, T2, T3, and WM) with relation to determined clusters of the PQ data.

The lack of a background color means that the data were flagged. It can be noticed that changeability of the *ADI* for different working conditions (represented by clusters) is observable but very faint. Thus, the results of the power quality assessment using the proposed technique combining the CA global power quality indices means that *ADI* can be summarized by statistics of the *ADI* variation for particular clusters and measurement points. The results are collected in Table 3.

Table 3. Results of the assessment of the power quality using the proposed global power quality indices *ADI* and *FDI* for the particular measurement points and with relation to clusters 1–3 when full definition of the *ADI* index is implemented.

Cluster		Cluster 1—DG Working				Cluster 2—DG Switched Off				Cluster 3—DG Switched Off and With a Different Network Topology Configuration			
Measurement Point		T1	T2	T3	WM	T1	T2	T3	WM	T1	T2	T3	WM
ADI	Minimal value	0.119	0.087	0.115	0.130	0.060	0.069	0.060	0.084	0.071	0.062	0.075	0.092
	Mean value	0.121	0.089	0.105	0.137	0.113	0.098	0.124	0.148	0.114	0.091	0.116	0.142
	Maximum value	0.123	0.092	0.115	0.145	0.259	0.188	0.256	0.260	0.212	0.382	0.178	0.203
FDI (%)		0.00				0.15				0.17			

Comparative analysis of the *ADI* levels allows the formulation of the following remarks:

- Transformers T2 and T3, as well as the connection point of the welding machine WM, had the highest level of *ADI* for cluster 2 when the DG was switched off, and the lowest for cluster 1 when the DG was active. Distributed generation units were connected directly to T2 and T3 and the impact of the DGs was identified.
- Transformer T1 had relatively higher *ADI* values for cluster 1 when the DG was active, and the lowest for cluster 2 when the DG was switched off. However, there was active generation directly connected to transformer T1.
- The highest level of *ADI* was recognized in the outcoming feeder that supplies the welding machine which is a significant load with highly time-varying nature.

- Referring to cluster 1 when the DG was active, the *ADI* had the lowest level for T2, then T3, and the highest for T1.
- Referring to cluster 2 when the DG was switched off, the *ADI* had the lowest level for T2, and the highest for T3.
- Cluster 3 represents a short period of time (around 2 days) when all the DGs were switched off and some reconfiguration of the electrical power network connection was made. During the reconfiguration, transformer T1 was more loaded, and transformers T2 and T3 were less loaded. Comparing the *ADI* level during cluster 3, consisting of a period of time when there was a network reconfiguration with cluster 2, when the network was operating in the normal configuration, it can be seen that the values of the *ADI* decreased for T2, T3 and WM, and increased for T1.

To sum up, using the proposed cluster analysis and the proposed global power quality index, the *ADI* can be a suitable tool for the identification and comparative assessment of different conditions of the observed network. We revealed that for the observed transformers T2, T3, and the connection point of welding machine WM, the power quality was better in cluster 1 when the DG was active. The different outcomes of the *ADI* level formulated for transformer T1 could be caused by the fact that there was no DG directly connected to T1. The highest values of *ADI* were identified in the feeder supplying the welding machine.

The next global power quality index proposed in this work is the flagged data index (*FDI*), which is related to the number of aggregated data affected by the events in reference to the periods identified by clusters. Comparative analysis of the *FDI* levels is presented in Table 3. It allows for the formulation of a general remark that the *FDI* level was noticeable for cluster 2 and cluster 3. The high values for cluster 2 and cluster 3 are probably connected with the events caused by changes in the electrical power network topology.

Additionally, correlation calculations between each factor to the *ADI* value were realized for each point separately. Pearson correlation was used and the description of the coefficient was defined as [10]:

- $r_{xy} = 0$—no correlation;
- $0 < |r_{xy}| \le 0.1$—slight correlation;
- $0.1 < |r_{xy}| \le 0.4$—poor correlation;
- $0.4 < |r_{xy}| \le 0.7$—noticeable correlation;
- $0.7 < |r_{xy}| \le 0.9$—high correlation;
- $0.9 < |r_{xy}|$—strong correlation.

The correlation between factors and *ADI* are presented in Table 4. The generally noticeable correlations in each measurement point were indicated for P_{st} (W_3), *THDu* (W_5), U_{env} (W_6), and THD_{Umax} (W_7).

Table 4. Results of correlation analysis between each factor and the global index.

Measurement Point	W_1	W_2	W_3	W_4	W_5	W_6	W_7
T1	slight	slight	high	slight	high	noticeable	high
T2	slight	poor	noticeable	poor	noticeable	noticeable	high
T3	poor	slight	noticeable	poor	high	high	high
WM	poor	poor	noticeable	poor	high	noticeable	high

3.3. *Influence of the Factors Comprising the Proposed Global Power Quality Indices on the Sensitivity of the Assessment*

The construction of a global power quality index, *ADI*, understood as a weighted sum of component factors related to power quality parameters, inclines us to discuss the impact of individual factors on the assessment results. It is possible to select weighting coefficients in a way that favors the selected parameters in the assessment and moves the center of gravity of the global assessment in the direction of the favorite parameters. The opposite direction is to enhance the sensitivity of the assessment by including additional parameters in the definition of global indices. This work proposes the construction of a global index using five basic 10-min parameters of power quality (frequency, root mean square (RMS) voltage, asymmetry, voltage fluctuations, and total harmonic distortion in voltage) and to extend the definition with two other parameters which are close to 200-ms values (i.e., the envelope of voltage changes and the maximum value of the total harmonic distortion in voltage identified during 10-min aggregation intervals). The aim of extending the *ADI* definition with parameters related to 200-ms intervals is to enhance the sensitivity of the obtained global index. In order to investigate the impact of the proposed 200-ms parameters on the sensitivity of the assessment, a differential approach is proposed. The *ADI* values for particular clusters and points of measurements when the full definition is involved are presented in Table 3. The results represent a scenario where all seven factors with the same weighting factors equal to $k_1 \div k_7 = 1/7$ were applied in the *ADI* calculation. Application of the full definition of *ADI* allowed us to conclude that for the observed transformers T2, T3, and the connection point of the welding machine WM, the power quality was better in cluster 1 when the DG was active. The obtained *ADI* values were generally smaller in cluster 1 than in cluster 2, and the differences of *ADI* between clusters 1 and 2 were consistently mostly negative. In order to perform a differential comparison between the *ADI* obtained using the full definition and the *ADI* based on a reduced definition, new values of the *ADI* were calculated where the parameters related to 200-ms values were neglected (i.e., when weighting factors were equal to $k_1 \div k_5 = 1/5$, $k_6 = 0$ and $k_7 = 0$, respectively). The obtained values of the *ADI* calculated without the 200-ms parameters are presented in Table 5.

Table 5. Results of the power quality assessment using the proposed global power quality index *ADI* for the selected measurement points, with relation to the revealed clusters when 200-ms parameters are neglected in the *ADI* definition.

Cluster		Cluster 1—DG Working				Cluster 2—DG Switched Off				Cluster 3—DG Switched Off and with a Different Network Topology Configuration			
Measurement point		T1	T2	T3	WM	T1	T2	T3	WM	T1	T2	T3	WM
ADI	Minimal value	0.106	0.095	0.099	0.124	0.052	0.064	0.056	0.090	0.071	0.062	0.075	0.092
	Mean value	0.110	0.095	0.097	0.133	0.099	0.099	0.111	0.144	0.114	0.091	0.116	0.142
	Maximum value	0.114	0.096	0.104	0.142	0.221	0.193	0.219	0.231	0.212	0.382	0.178	0.203

Instead of calculating the direct differences between the *ADI* values obtained for both scenarios (which actually differ very slightly), we propose a comparison between interpretations of the results. In other words, the sensitivity analysis was redirected to formulate the question of whether neglecting the 200-ms parameters in the *ADI* definition would change the interpretation of the assessment. Changes in the interpretation of the results can be identified if the signs of the difference of the *ADI*s applied for full and reduced definitions are different. For example, we found that the *ADI* obtained using the full definition decreased when the DG is active (C1—cluster 1) and increased when the DG was switched off (C2—cluster 2). The difference of the *ADI*s between C1 and C2 was negative because the values of the *ADI* in C2 predominated. If a reduction of the *ADI* parameters has an influence on the sign of the differences between the clusters, it means that the interpretation is not coherent and is dependent on the *ADI* construction. Table 6 contains information about the assessment results between the *ADI* with 200-ms factors ($k_1 \div k_7 = 1/7$) and without 200-ms factors ($k_1 \div k_5 = 1/5$, $k_6 = 0$, and $k_7 = 0$). Additionally, an interpretative logical comparative assessment index is introduced in the table. A value equal to 1 means that the assessment and interpretation of the results are the same for the full and reduced definitions of the *ADI*. A value equal to −1 means that the interpretations using full and reduced definitions of the *ADI* are not coherent.

Table 6. Assessment of the influence of removing two parameters related to 200-ms intervals in the *ADI* definition on the power quality assessment. The interpretation differences were highlighted.

ADI	Delta of Clusters	T1			T2			T3			WM		
		Min	Mean	Max	Min	Mean	Max	Min	Mean	Max	Min	Mean	Max
with 200 ms	Δ C1–C2	0.059	0.008	−0.136	0.017	−0.009	−0.097	0.055	−0.019	−0.141	0.046	−0.010	−0.114
	Δ C1–C3	0.048	0.007	−0.089	0.025	−0.001	−0.290	0.040	−0.012	−0.063	0.038	−0.004	−0.058
	Δ C2–C3	−0.012	−0.001	0.047	0.007	0.007	−0.193	−0.014	0.008	0.078	−0.008	0.006	0.056
without 200 ms	Δ C1–C2	0.053	0.011	−0.107	0.030	−0.004	−0.097	0.043	−0.014	−0.115	0.034	−0.011	−0.089
	Δ C1–C3	0.034	−0.004	−0.097	0.033	0.005	−0.286	0.024	−0.020	−0.075	0.032	−0.009	−0.061
	Δ C2–C3	−0.019	−0.015	0.010	0.002	0.009	−0.188	−0.019	−0.006	0.041	−0.002	0.002	0.028
Logical comparative assessment		1	1	1	1	1	1	1	1	1	1	1	1
no change in assessment: 1		1	−1	1	1	−1	1	1	1	1	1	1	1
change in assessment: −1		1	1	1	1	1	1	1	−1	1	1	1	1

Based on the analysis presented in Table 6, it can be generally concluded that among the 36 assessments of the clusters, 3 differ in terms of the interpretation after a reduction of the *ADI* definition. In other words, a reduction of the *ADI* components introduced an 8% difference in the assessment. Alternatively, this means that including the parameters associated with the 200-ms values in the *ADI* definition enhances the sensitivity of the assessment.

To be more precise, the comparison of the differences of the *ADI* values constructed on seven and five parameters addressed to particular clusters were seen to deliver additional observations. For clusters 1 and 2, it can be concluded that interpretation results based on the *ADI*s were not sensitive to a reduction of *ADI* components, and the interpretation results were the same. This is due to the substantial differences between the power quality condition in clusters 1 and 2, which are reflected in the *ADI* values. However, when comparing clusters containing similar data, the reduction of the *ADI* components may cause differences in the assessment. An example of this can be seen with the data associated with transformer T3 in cluster 2 (DG switched off) and cluster 3 (DG switched off and with network reconfiguration). In this case, there was a significant impact of the DG; the power quality conditions were similar, and the reduction of *ADI* components brought differences in the interpretation in Table 6. This is denoted by the logical value −1. Another example can be seen in the differences between cluster 1 and cluster 3 in the case of transformer T2. The network configuration result was more loaded in transformer T1 and less loaded in transformer T2. In terms of transformer T2, it was generally a similar condition as for the impact of DG when the reduction of the load demand was also achieved. In this case, the power quality condition was similar for both clusters 1 and 3, and the reduction of the *ADI* components introduced uncertainty to the assessment.

It can be concluded that the reduction of the parameters comprising the synthetic *ADI* index influences the sensitivity of the assessment. In the case of the presented investigation, this inherent relation was more significant when the differences between the power conditions in the compared clusters were insignificant.

4. Discussion

This work presents the possibility of connecting CA and GPQIs. As indicated by the authors in a previous work [9] PQ measurements are an appropriate input to cluster analysis. Note that the aim of CA is to divide data based on its features. The proposed method was implemented for the real measurements collected from four measurement points in an industry network: three transformers T1, T2, T3 which supplied the MV industrial network, and a significant load (a welding machine, WM). The investigation aimed to evaluate the influence of the DGs installed inside the observed industrial network. However the power variations of the DGs are additional information, representing conditions. The data from the DGs do not create the database of measurements taken for the investigations. Naturally the same classification can be obtained using time identification representing different conditions of the DGs, but the point of the method is to obtain automatic classification of the PQ data based on its features, and then to find the reasons explaining the automatic classification. The presented approach has a crucial meaning when the number of monitored points is increased.

The input database consists of many different parameters, leading to a multielement assessment. Thus, in this work we proposed the use of global indices to simplify the process. The proposed indices consist of power quality parameters that represent frequency, voltage, flicker asymmetry factor, and harmonics in voltage. To classical 10-min aggregated data, we proposed adding the extremum values of voltage and harmonics. Thus, we conducted an analysis of the impact of extending the global indices to such values. Results indicated that our synthetic *ADI* index influenced the sensitivity of the assessment. In the case of the present investigation, this inherent relation was more significant when the differences between the power conditions in the compared clusters were insignificant. The composition of *ADI* index is based on classical 10-min PQ parameters as well as 200-ms parameters. Weighting factors were implemented for particular parameters. In order to reveal the influence of the DGs, all weighting factors were set to one in order to obtain maximum sensitivity of the analysis on every PQ parameter collected in *ADI*. However, the weighted factors make it possible to focus the analysis more on particular PQ parameters and neglect others (i.e., to obtain an analysis more sensitive for selected PQ phenomena controlled by using different values of the weighting factor).

The proposed combination of CA and GPQIs was indicated as a suitable tool for the identification and comparative assessment of different conditions of the observed mining industry network. Among other things it was revealed that for the observed transformers T2, T3, and the connection point of welding machine WM, the power quality was better in cluster 1 when the DG was active. The different outcomes of the *ADI* level for transformer T1 could be caused by the fact that there was no DG directly connected to T1. The highest values of *ADI* were identified in the feeder supplying the welding machine, which is a high variable load. It can be concluded that obtained method is also technically reasonable.

We also proposed the flagged data index (*FDI*), which is related to the number of aggregated data affected by the events. It was used to compare clusters. Results concerning the use of the proposed global power quality index dedicated to voltage events (*FDI*) showed that the *FDI* was higher in cluster 2 than in cluster 1, which can be attributed to the fact that in the period of time when DG was active (cluster C1) there was relatively fewer detected voltage events than in the period when the DG was switched off (cluster 2). The sense of the *FDI* is general. Detailed analysis of particular voltage events requires separate investigations.

5. Conclusions

This article presents a method of analyzing long-term PQ data using a combined technique based on cluster analysis and newly proposed global power quality indices. The presented investigations were based on multipoint synchronized real measurements performed in a medium voltage electrical power network with distributed generation supplying the mining industry. Time-varying PQ conditions were intentionally created during the experiment when the distributed generation was switched on and off for some period of time, with a network reconfiguration also being performed.

The cluster analysis is the first step of the proposed method and is used for identification of the PQ data which represent different conditions. It was shown that cluster analysis with K-means and Euclidean distance successfully allowed for the identification of portions of PQ data related to the impact of distributed generation (switched on and switched off) and changes to the network configuration. Basic investigations of the application of cluster analysis in an electrical power network were presented by the authors in a previous work [9]. The extension of the mentioned work and the novelty involved in the proposed method lies in extending the cluster analysis by assessing the obtained portions of PQ data using global power quality indices. In order to achieve the goal, newly proposed global power quality indices were provided, including the aggregated data index and flagged data index. The proposed aggregated data index has a synthetic formula and is based on five classical 10-min aggregated power quality parameters and two parameters that demonstrate 200-ms values, including the envelope of voltage changes and the maximum of total harmonic distortion in the voltage. In this work, the proposed global indices were used for comparative assessment of identified clusters, which in turn demonstrated different states of the network condition: active distributed generation, switched off generation, and network reconfiguration when the generation was switched off. It was shown that the use of the proposed global power quality indices resulted in the comparative analysis between particular clusters being successfully performed.

Additionally, a sensitivity analysis of the synthetic aggregation data index was also proposed. It can be concluded that a reduction of the parameters comprising the synthetic global power quality index may influence the results of the assessment. In the case of the presented investigation, this inherent relation was more significant when the differences between power conditions in the compared clusters were insignificant.

The presented approach can be treated as an effective tool (not only related to power quality) for the assessment of long-term multipoint measurements. The advantages of the proposed method are the automatic classification of the data into clusters and the assessment of the condition of the identified group of data in a parametric global sense, which makes the comparative assessment easier and more intuitive. The proposed technique has the potential for further implementation in the analysis and optimization of energy processes, and also in the development of sustainable energy systems.

Author Contributions: Conceptualization, M.J. and T.S.; methodology, M.J. and T.S.; software, M.J.; validation, K.B.; formal analysis, M.J.; investigation, M.J.; resources, P.K. and K.B.; data curation, P.K.; writing—original draft preparation, M.J.; writing—review and editing, T.S.; visualization, M.J.; supervision, T.S. and Z.L.; project administration, T.S.; funding acquisition, Z.L. All authors have read and agreed to the published version of the manuscript.

Funding: This research received funding from the Chair of Electrical Engineering Fundamentals (K38W05D02), Wroclaw University of Technology, Wroclaw, Poland.

Acknowledgments: The authors also acknowledge the support of KGHM Polska Miedz SA.

Conflicts of Interest: The authors declare no conflicts of interest.

Nomenclature:

ADI	aggregated data index
C	database for non-standardized data
Cs	database for standardized data
C	number of classes or clusters
DG	distributed generation
GPQI	global power quality index
k_i	importance rate (weighted factors) of a particular power quality factor constituting the synthetic aggregated data index, range of [0, 1]
k_{u2}	asymmetry
P	active power
P_{lt}	long-term flicker severity
P_{st}	short-term flicker severity
PQ	power quality
THD	total harmonic distortion
U	voltage variation
W_i	particular power quality factors comprising the synthetic aggregated data index
WM	welding machine

References

1. International Energy Agency. *Key World Energy Statistics*; International Energy Agency: Paris, France, 2017.
2. Nieto, A.; Vita, V.; Ekonomou, L.; Mastorakis, N.E. Economic Analysis of Energy Storage System Integration with a Grid Connected Intermittent Power Plant, for Power Quality Purposes. *WSEAS Trans. Power Syst.* **2016**, *11*, 65–71.
3. Nieto, A.; Vita, V.; Maris, T.I. Power Quality Improvement in Power Grids with the Integration of Energy Storage Systems. *Int. J. Eng. Res. Technol.* **2016**, *5*, 438–443.
4. Ceaki, O.; Seritan, G.; Vatu, R.; Mancasi, M. Analysis of power quality improvement in smart grids. In Proceedings of the 2017 10th International Symposium on Advanced Topics in Electrical Engineering (ATEE), Bucharest, Romania, 23–25 March 2017; IEEE: Piscataway, NJ, USA, 2017; pp. 797–801.
5. Seritan, G.; Tristiu, I.; Ceaki, O.; Boboc, T. Power quality assessment for microgrid scenarios. In Proceedings of the 2016 International Conference and Exposition on Electrical and Power Engineering (EPE), Iasi, Romania, 20–22 October 2016; IEEE: Piscataway, NJ, USA, 2016; pp. 723–727.
6. *EN 50160: Voltage Characteristics of Electricity Supplied by Public Distribution Network*; British Standards: London, UK, 2015.
7. *IEC 61000 4-30 Electromagnetic Compatibility (EMC)–Part 4-30: Testing and Measurement Techniques–Power Quality Measurement Methods*; International Electrotechnical Commission: Geneva, Switzerland, 2015.
8. *IEEE P1159: Monitoring and Definition of Electric Power Quality*; IEEE: Piscataway, NJ, USA, 2009.

9. Jasiński, M.; Sikorski, T.; Borkowski, K. Clustering as a tool to support the assessment of power quality in electrical power networks with distributed generation in the mining industry. *Electr. Power Syst. Res.* **2019**, *166*, 52–60. [CrossRef]
10. Jasiński, M.; Sikorski, T.; Kostyła, P.; Kaczorowska, D.; Leonowicz, Z.; Rezmer, J.; Szymańda, J.; Janik, P.; Bejmert, D.; Rybiański, M.; et al. Influence of Measurement Aggregation Algorithms on Power Quality Assessment and Correlation Analysis in Electrical Power Network with PV Power Plant. *Energies* **2019**, *12*, 3547. [CrossRef]
11. Borges, F.A.S.; Fernandes, R.A.S.; Silva, I.N.; Silva, C.B.S. Feature Extraction and Power Quality Disturbances Classification Using Smart Meters Signals. *IEEE Trans. Ind. Inform.* **2015**, *12*, 824–833. [CrossRef]
12. Erişti, H.; Demir, Y. Automatic classification of power quality events and disturbances using wavelet transform and support vector machines. *IET Gener. Transm. Distrib.* **2012**, *6*, 968–976. [CrossRef]
13. Jamali, S.; Farsa, A.R.; Ghaffarzadeh, N. Identification of optimal features for fast and accurate classification of power quality disturbances. *Meas. J. Int. Meas. Confed.* **2018**, *116*, 565–574. [CrossRef]
14. CIGRE. *Broshure 292: Data Mining Techniques and Applications in the Power Transmission Field*; CIGRE: Paris, France, 2006.
15. Xi, Y.; Li, Z.; Zeng, X.; Tang, X.; Liu, Q.; Xiao, H. Detection of power quality disturbances using an adaptive process noise covariance Kalman filter. *Digit. Signal Process. A Rev. J.* **2018**, *76*, 34–49. [CrossRef]
16. Kanirajan, P.; Suresh Kumar, V. Power quality disturbance detection and classification using wavelet and RBFNN. *Appl. Soft Comput. J.* **2015**, *35*, 470–481. [CrossRef]
17. Mahela, O.P.; Shaik, A.G. Power quality recognition in distribution system with solar energy penetration using S -transform and Fuzzy C-means clustering. *Renew. Energy* **2017**, *106*, 37–51. [CrossRef]
18. Gu, T.; Kadurek, P.; Cobben, J.F.G.; Endhoven, A.W. Power quality data evaluation in distribution networks based on data mining techniques. In Proceedings of the 2013 12th International Conference on Environment and Electrical Engineering, Wroclaw, Poland, 5–8 May 2013; IEEE: Piscataway, NJ, USA, 2013; pp. 58–63.
19. Gao, Q.; Pan, F.; Yuan, F.; Pan, J.; Zhang, J.; Zhang, Y. The Classification of Multiple Power Quality Disturbances Based on Dynamic Event Tree and Support Vector Machine. In *International Conference on Advanced Machine Learning Technologies and Applications*; Springer: Cham, Switzerland, 2020; pp. 321–329.
20. Bagheri, A.; Bollen, M.H.J.; Gu, I.Y.H. Improved characterization of multi-stage voltage dips based on the space phasor model. *Electr. Power Syst. Res.* **2018**, *154*, 319–328. [CrossRef]
21. Ahire, P.G.; Patil, P.D.; Patil, R.A.; Golande, A.L. An Approach to Clustering of Power Quality Events Data. *Proc. Int. J. Eng. Innov. Technol.* **2012**, *2*, 164–168.
22. Mittal, D.; Mahela, O.P.; Jain, R. Classification of Power Quality Disturbances in Electric Power System: A Review. *Iosr-Jeee* **2012**, *3*, 6–14. [CrossRef]
23. Shaik, A.G.; Mahela, O.P. Power quality assessment and event detection in hybrid power system. *Electr. Power Syst. Res.* **2018**, *161*, 26–44. [CrossRef]
24. Meier, R.; McCamish, B.; Cotilla-Sanchez, E.; Landford, J.; Bass, R.B.; Chiu, D. Event Detection Using Correlation within Arrays of Streaming PMU Data. In Proceedings of the 2018 IEEE Power & Energy Society General Meeting (PESGM), Portland, OR, USA, 5–10 August 2018; IEEE: Piscataway, NJ, USA, 2018; pp. 1–5.
25. Guder, M.; Salor, O.; Çadirci, N.; Ozkan, B.; Altintas, A. Data Mining Framework for Power Quality Event Characterization of Iron and Steel Plants. *Proc. Ind. Appl. Soc. Annu. Meet* **2015**, *51*, 3521–3531. [CrossRef]
26. Dent, I.; Aickelin, U.; Rodden, T. The Application of a Data Mining Framework to Energy Usage Profiling in Domestic Residences Using UK Data. *SSRN Electron. J.* **2011**, 1–10. [CrossRef]
27. Asheibi, A.; Stirling, D.; Robinson, D. Identification of Load Power Quality Characteristics using Data Mining. In Proceedings of the 2006 Canadian Conference on Electrical and Computer Engineering, Ottawa, ON, Canada, 7–10 May 2006; IEEE: Piscataway, NJ, USA, 2006; pp. 157–162.
28. Zhou, K.; Yang, S.; Shen, C. A review of electric load classification in smart grid environment. *Renew. Sustain. Energy Rev.* **2013**, *24*, 103–110. [CrossRef]
29. Alfares, H.K.; Nazeeruddin, M. Electric load forecasting: Literature survey and classification of methods. *Int. J. Syst. Sci.* **2002**, *33*, 23–34. [CrossRef]
30. Kotriwala, A.M.; Hernandez-Leal, P.; Kaisers, M. Load Classification and Forecasting for Temporary Power Installations. In Proceedings of the 2018 IEEE PES Innovative Smart Grid Technologies Conference Europe (ISGT-Europe), Sarajevo, Bosnia-Herzegovina, 21–25 October 2018; IEEE: Piscataway, NJ, USA, 2018; pp. 1–6.

31. Azizi, E.; Ghaemi, S.; Mohammadi-Ivatloo, B.; Piran, M.J. Application of comparative strainer clustering as a novel method of high volume of data clustering to optimal power flow problem. *Int. J. Electr. Power Energy Syst.* **2019**, *113*, 362–371. [CrossRef]
32. Stankovic, S.; Soder, L. Optimal Power Flow Based on Genetic Algorithms and Clustering Techniques. In Proceedings of the 2018 Power Systems Computation Conference (PSCC), Dublin, Ireland, 11–15 June 2018; IEEE: Piscataway, NJ, USA, 2018; pp. 1–7.
33. Huynh, V.; Ngo, V.; Le, D.; Nguyen, N. Probabilistic Power Flow Methodology for Large-Scale Power Systems Incorporating Renewable Energy Sources. *Energies* **2018**, *11*, 2624. [CrossRef]
34. Galvani, S.; Choogan, M. Data clustering based probabilistic optimal power flow in power systems. *IET Gener. Transm. Distrib.* **2019**, *13*, 181–188. [CrossRef]
35. Florencias-Oliveros, O.; Aguera-Perez, A.; Gonzalez-De-La-Rosa, J.J.; Palomares-Salas, J.C.; Sierra-Fernandez, J.M.; Montero, A.J. Cluster analysis for Power Quality monitoring. In Proceedings of the 2017 11th IEEE International Conference on Compatibility, Power Electronics and Power Engineering (CPE-POWERENG), Cadiz, Spain, 4–6 April 2017.
36. Lin, S.; Xie, C.; Tang, B.; Liu, R.; Pan, A.; Grid, S.; Municipal, S.; Power, E. The Data Mining Application in the Power Quality Monitoring Data Analysis. In Proceedings of the 2016 IEEE 11th Conference on Industrial Electronics and Applications (ICIEA), Hefei, China, 5–7 June 2016; pp. 338–342.
37. Seera, M.; Lim, C.P.; Loo, C.K.; Singh, H. A modified fuzzy min-max neural network for data clustering and its application to power quality monitoring. *Appl. Soft Comput. J.* **2015**, *28*, 19–29. [CrossRef]
38. Seera, M.; Lim, C.P.; Loo, C.K.; Singh, H. Power Quality Analysis Using a Hybrid Model of the Fuzzy Min–Max Neural Network and Clustering Tree. *IEEE Trans. Neural Networks Learn. Syst.* **2016**, *27*, 2760–2767. [CrossRef] [PubMed]
39. Sacasqui, M.; Luyo, J.; Delgado, A. A Unified Index for Power Quality Assessment in Distributed Generation Systems Using Grey Clustering and Entropy Weight. In Proceedings of the 2018 IEEE ANDESCON, Santiago de Cali, Colombia, 22–24 August 2018; IEEE: Piscataway, NJ, USA, 2018; pp. 1–4.
40. Elbasuony, G.S.; Abdel Aleem, S.H.E.; Ibrahim, A.M.; Sharaf, A.M. A unified index for power quality evaluation in distributed generation systems. *Energy* **2018**, *149*, 607–622. [CrossRef]
41. Song, J.; Xie, Z.; Zhou, J.; Yang, X.; Pan, A. Power quality indexes prediction based on cluster analysis and support vector machine. *CIRED-Open Access Proc. J.* **2017**, *2017*, 814–817. [CrossRef]
42. Florencias-Oliveros, O.; Agüera-Pérez, A.; González-de-la-Rosa, J.J.; Palomares-Salas, J.-C.; Sierra-Fernández, J.-M. A novel instrument for power quality monitoring based in higher-order statistics: A dynamic triggering index for the smart grid. *Renew. Energy Power Qual. J.* **2017**, *1*, 43–48. [CrossRef]
43. Singh, L.P.; Deswal, S.S.; Bhatia, R.; Jain, D.K. A New Approach for Analysing Voltage Sag Severity Based on Power Quality Indices. *Int. J. Emerg. Technol.* **2004**, *4*, 1–5.
44. Ignatova, V.; Villard, D.; Hypolite, J.M. Simple indicators for an effective Power Quality monitoring and analysis. In Proceedings of the 2015 IEEE 15th International Conference on Environment and Electrical Engineering, EEEIC 2015-Conference Proceedings, Rome, Italy, 10–13 June 2015; pp. 1104–1108.
45. Nourollah, S.; Moallem, M. A Data Mining Method for Obtaining Global Power Quality Index. In Proceedings of the 2011 2nd International Conference on Electric Power and Energy Conversion Systems (EPECS), Sharjah, United Arab Emirates, 15–17 November 2011; pp. 1–7.
46. Raptis, T.E.; Vokas, G.A.; Langouranis, P.A.; Kaminaris, S.D. Total Power Quality Index for Electrical Networks Using Neural Networks. *Energy Procedia* **2015**, *74*, 1499–1507. [CrossRef]
47. Lee, S.; Park, J.; Member, S.; Venayagamoorthy, G.K. New Power Quality Index in a Distribution Power System by Using RMP Model. *IEEE Trans. Ind. Appl.* **2010**, *46*, 1204–1211.
48. Hanzelka, Z.; Firlit, A.; Błajszczak, A. Syntetyczne miary jakości napięcia. *Autom. Elektr. Zakłócenia* **2012**, *3*, 62–70.
49. Caramia, P.; Carpinelli, G.; Verde, P. *Power Quality Indices in Liberalized Markets.pdf*; John Wiley & Sons: Hoboken, NJ, USA, 2009; ISBN 9780470033951.
50. Jasiński, M.; Sikorski, T.; Kostyla, P.; Borkowski, K. Global power quality indices for assessment of multipoint power quality measurements. In Proceedings of the Electronics, Computers and Artificial Intelligence, Iasi, Romania, 28–30 June 2018; IEEE: Piscataway, NJ, USA, 2018; pp. 1–6.

51. Jasiński, M.; Borkowski, K.; Sikorski, T.; Kostyla, P. Cluster Analysis for Long-Term Power Quality Data in Mining Electrical Power Network. In Proceedings of the 2018 Progress in Applied Electrical Engineering (PAEE), Koscielisko, Poland, 18–22 June 2018; IEEE: Piscataway, NJ, USA, 2018; pp. 1–5.
52. Jasiński, M.; Sikorski, T.; Borkowski, K. Application of cluster analysis to identification flagged power quality measurements in area-related approach. [Zastosowanie eksploracji danych do identyfikacji oznaczonych wyników pomiaru jakosci energii elektrycznej w ujeciu obszarowym]. *Prz. Elektrotechniczny* **2020**, *3*, 9–12.

energies

MDPI

Article

The Application of Hierarchical Clustering to Power Quality Measurements in an Electrical Power Network with Distributed Generation

Michał Jasiński [1,*], Tomasz Sikorski [1], Zbigniew Leonowicz [1], Klaudiusz Borkowski [2] and Elżbieta Jasińska [3]

1 Department of Electrical Engineering Fundamentals, Faculty of Electrical Engineering, Wroclaw University of Science and Technology, 50-370 Wroclaw, Poland
2 KGHM Polska Miedź S.A., 50-301 Lubin, Poland
3 Faculty of Law, Administration and Economics, University of Wroclaw, 50-145 Wroclaw, Poland
* Correspondence: michal.jasinski@pwr.edu.pl; Tel.: +48-713-202-022

Received: 23 April 2020; Accepted: 9 May 2020; Published: 11 May 2020

Abstract: This article presents the application of data mining (DM) to long-term power quality (PQ) measurements. The Ward algorithm was selected as the cluster analysis (CA) technique to achieve an automatic division of the PQ measurement data. The measurements were conducted in an electrical power network (EPN) of the mining industry with distributed generation (DG). The obtained results indicate that the application of the Ward algorithm to PQ data assures the division with regards to the work of the distributed generation, and also to other important working conditions (e.g., reconfiguration or high harmonic pollution). The presented analysis is conducted for the area-related approach—all measurement point data are connected at an initial stage. The importance rate was proposed in order to indicate the parameters that have a high impact on the classification of the data. Another element of the article was the reduction of the size of the input database. The reduction of input data by 57% assured the classification with a 95% agreement when compared to the complete database classification.

Keywords: data mining; power quality; cluster analysis; ward algorithm; different working conditions; distributed generation

---

1. Introduction

A smart grid can be seen as the future of electrical power systems [1–3]. A smart grid requires the monitoring and cooperation of more and more elements, devices, and systems. Thus, it introduces the need for analyzing an increasing amount of data. Single parameter analysis, conducted by humans, has become a thing of the past in terms of the functioning of an electrical power system (EPS). Thus, a need for tools to support the long-term assessment has become very necessary [4–7].

This research is a continuation of previous work [8], which involves a method for analyzing long-term power quality (PQ) data using non-hierarchical clustering and its assessment using global indices in [9]. The presented results in Jasiński et al. [8] were based on 72 cases of clustering, which differ in terms of both the number of clusters (2/25), and also the distance definition of the items in the database (Euclidean, Chebyshev) for the K-mean algorithm. The different constructions of the database were discussed. The direct impact of the distributed generation (DG) on the PQ conditions was obtained when clustering using the K-mean algorithm and the Euclidean distance for non-standardized data that are extended by power consumption, using database C: frequency (f), voltage variations (U), short term flicker severity (P_{st}), asymmetry (k_{u2}), total harmonic distortion in voltage (THD_U), active power (P). Thus, in this article, the same input database was selected. However, the Ward algorithm is presented in this research, which represents the hierarchical approach. Additionally, this work contains an analysis of the importance rate in order to indicate which parameters have an impact on the final classification. The comparison of clusters, which represent different working conditions of the electrical power network (EPN), obtained automatically, was only conducted for the indicated parameters with a high importance rate but not using a global index, which includes all the parameters as in [9]. Additionally, the next novelty of this work is the proposition of reducing the input database without losing data features. The proposed reduction to one value, instead of three phase-to-phase parameters, assured the classification with a 95% agreement when compared to the complete database classification.

The article is organized into four sections. Section 2 presents the state of the art of literature review. Section 3 describes the definitions and techniques of cluster analysis (CA), with special consideration for the Ward algorithm. Also, Section 3 contains the description of the research object—the EPN of the mining industry with gas-steam units and conducted long-term PQ measurements. Additionally, Section 3 contains the application of the Ward algorithm to PQ data and the results of the analysis with regards to the different working conditions of the EPN. The final element of Section 3 presents a discussion of the obtained results. Section 4 highlights the conclusions.

2. Literate Review

One solution to the problem of big data analysis is the application of data mining (DM) techniques. The literature contains many examples of the possible applications of DM for electric power systems, e.g.,

- the detection and classification of voltage events [10–15]
- the calculation and prediction of power losses [16–18]
- the diagnosis of faults in power transformers [19–23]
- load forecasting [24–29]
- load pattern segmentation [30–33]
- fault detection [34–38]
- fault prediction [39–42]
- the defining of energy consumption [43–48]
- the forecasting of energy gaining from renewable energy sources [49–52]
- the reliability assessment of renewable sources of energy [53–56]
- energy management in a household [57–60]
- the improvement of intrusion detection systems in smart grids [61–63]
- the detection of electricity theft in smart grids [64–67]

As observed above, the application of data mining is wide. This article presents the application of data mining for achieving an automatic classification of long-term power quality (PQ) data from an electrical power network (EPN) of the mining industry with distributed generation (DG). The selected technique is cluster analysis (CA).

3. Methods and Results

3.1. Cluster Analysis—Ward Algorithm

Generally, the definition of data mining in the literature concerns the achievement of knowledge from big databases. Possible algorithms and techniques are well-known and described in the literature. Examples of data mining techniques are [68–72]:

- decision trees
- neural networks
- clustering

- regression
- mining association rules
- the multilayered perceptron network—MLP network
- genetic algorithms
- fuzzy interference systems
- high-performance computing
- inductive logic programming
- memory-based reasoning methods
- fuzzy sets

One of the described techniques is cluster analysis, also known as clustering [73]. The main aim of cluster analysis is to achieve homogeneous groups (clusters) of data as defined by Witten et al. and Wu et al. in [74,75]. The homogeneous aspect of the group is defined by the similarity or dissimilarity level of the data in the same cluster. There are a lot of data similarity/dissimilarity conditions that can be selected. However, due to the grouping process approach, two basic methods of dividing are known:

- hierarchical
- non-hierarchical

In this article, the hierarchical method is presented. Hierarchical approaches are agglomeration or divisive techniques. This article presents the agglomerative approach. Agglomerative techniques represent a set of observations in which each piece of data is treated as a separate cluster at the beginning. Then, the data are aggregated into a smaller number of clusters until one single cluster is established, which represents all the data [73]. The possible methods for connecting data into clusters are [73,76]:

- the single linkage method
- the complete linkage method
- the average linkage method
- the weighted pair-group average linkage method
- the unweighted pair-group centroid linkage method
- the unweighted pair-group centroid linkage method
- the Ward method of minimum variance

The hierarchical method is selected because the agglomerative sequence is presented on a dendrogram. It is, therefore, possible to analyze if the connection is better realized by single data or by a group of similar data (achieved in the previous agglomeration) to get a final classification. The authors selected the Ward algorithm due to its features. Clustering is carried out in order to connect data concentrated in an average value until the data has a similar value (range). The hierarchical cluster analysis algorithm using the Ward method of minimal variance is presented in Figure 1.

In this paper, the hierarchical Ward method and non-hierarchical method based on the K-mean algorithm are proposed for the power quality data analysis. The indicated "finding pair of clusters which have the smallest sum of squares distance between the object and the cluster center to which this object belongs", is calculated as presented in Equation (1) [77].

$$D_{pr} = \frac{n_p + n_r}{n_p + n_q + n_r} * d_{pr} + \frac{n_q + n_r}{n_p + n_q + n_r} * d_{qr} + \frac{-n_r}{n_p + n_q + n_r} * d_{pq} \tag{1}$$

where:

D_{pr}—distance of the new cluster to cluster of number "r",
r—proceed numbers of cluster from "p" to "q",
d_{pr}—distance of primary cluster "p" from cluster "r",

d_{qr}—distance of primary cluster "q" from cluster "r",
d_{pq}—common distance of primary clusters "p" and "q",
n—number of single objects in each object.

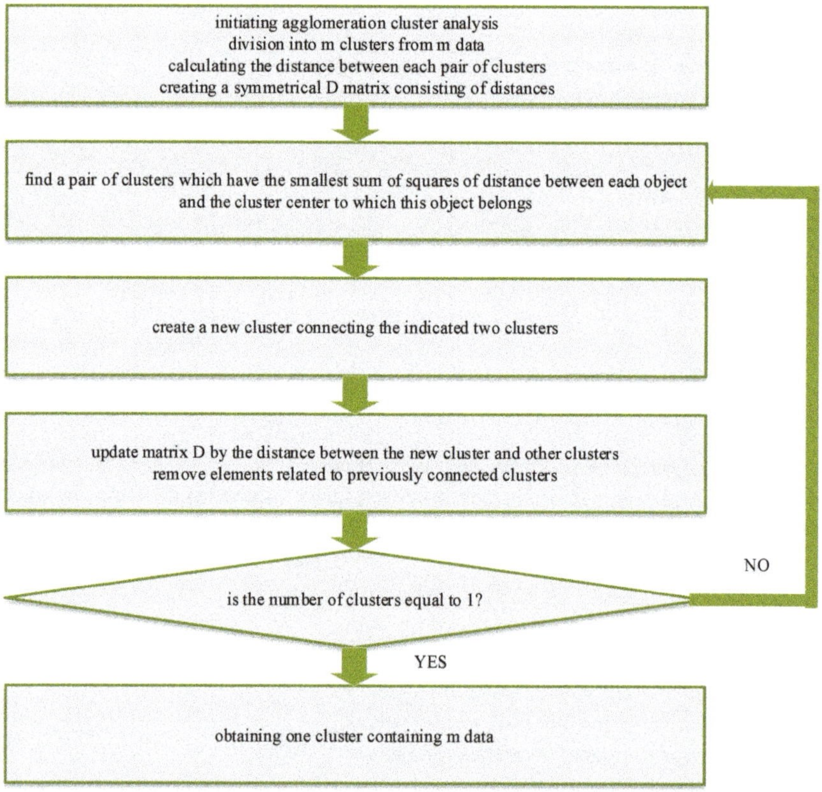

Figure 1. Cluster analysis using the Ward method of minimum variance [73,78].

Additionally, the advantage of the Ward algorithm is that it can be stopped at any moment; it can also achieve a classification represented by the excepted number of clusters. Thus, the final number of clusters should be selected in accordance with the aim of the classification. In order to support the final number of clusters, a lot of approaches have been conducted in literature. The most known are [79]:

- a dendrogram is analyzed in terms of the difference in distance between successive clusters. A big value of difference means that the data in the cluster are various. Thus, the division ends when the difference in the distance is maximal
- if a clear flattening (log vertical line) can be observed on the dendrogram, it means that in this point the clusters are distant and it is the best point for division
- an approach based on the root-mean-square standard deviation

3.2. *An Electrical Power Network of the Mining Industry and the Source of the PQ Data*

The PQ data used in the investigation concerns real measurements made in substations of the copper industry's electrical power network. The 110-kV substation of the mining industry works in a four-section system in cooperation with the four transformers (T1, T2, T3, T4). Normally, all the transformers are supplied from a different 110 kV section. However, during the measurements, the T4 transformer was not loaded. Thus:

- substations R-1 work independently
- substations R-2 work independently
- substations R-3 and R-4 are coupled

The presented PQ data concerns four weeks of measurements from 27th of April to 25th of May. The measurements were conducted synchronously with class A PQ recorders [80]. This is more than the classical one week of observation time, and therefore, the PQ data may consist of different working conditions of the analyzed electrical power network of the mining industry [77]. Thus, the different working conditions may be connected to:

MAIN LOADS:

- welding machines
- conveyor belts
- drainage pumps

DISTRIBUTED GENERATION:

- combined heat and power (CHP)
- gas-steam units

Thus, the PQ measurements include the analysis of the PQ level, which concerns the impact of the DG and main load (welding machine) on the medium voltage (MV) network. The simplified scheme of the copper industry network, showing the localization of power quality recorders installed in selected bays and the localization of DG, is presented in Figure 2. The PQ recorders involve the measurements of transformers at 6 kV side (T1, T2, T3) and an outcoming feeder to a welding machine (WM).

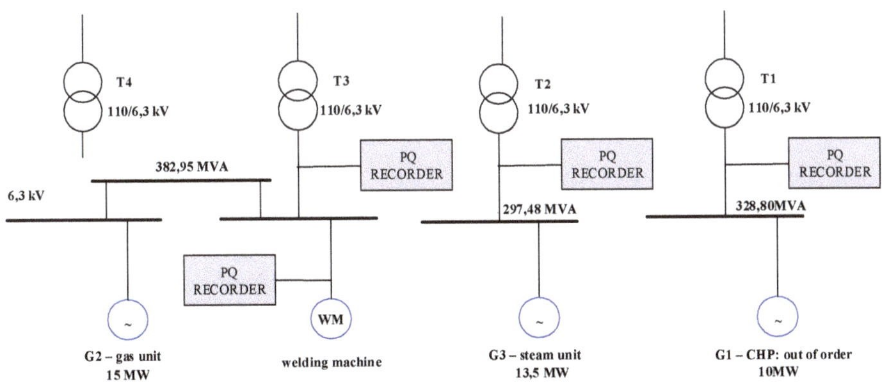

Figure 2. The simplified scheme of the electrical power network of the mining industry containing the placement of the PQ recorder and distributed generation.

It is important to note that the local generation is connected at the 6 kV level and that it consists of heat, a powerplant (G1–10 MW CHP), and steam-gas generation units (G3–15 MW gas unit and G2–13,5 MW steam unit). During the measurements, G1 was out of order and the level of generation of G2 and G3 was changing. The level of DG power (G1, G2, G3) and active power transformers (T1, T2, T3, WM) at the MV level are presented in Figure 3.

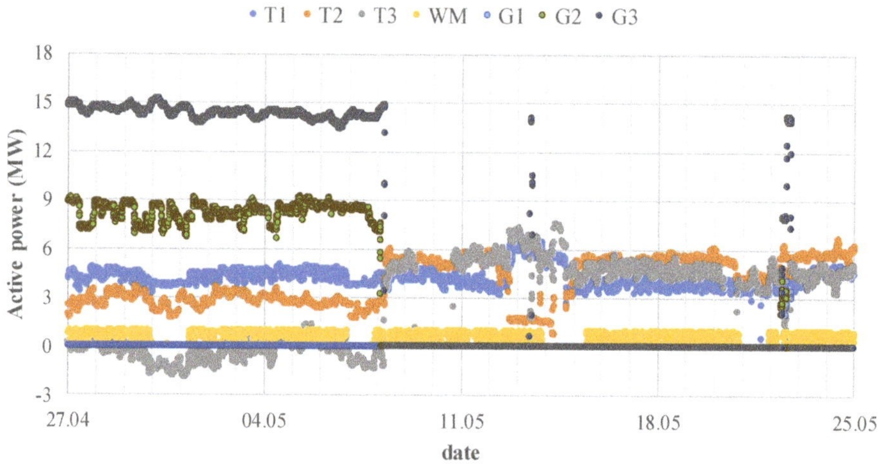

Figure 3. Active power of the high/medium (HV/MV) transformers T1, T2, and T3, welding machine (WM), and distributed generations G1, G2, G3 using the MV side measurements.

3.3. *Cluster Analysis Results*

3.3.1. Parameters Included to the Input Database

For the implementation of hierarchical cluster analysis, the Ward algorithm was used. The reason for this is due to the fact that the data assigned to clusters are characterized by the smallest variation of results (minimum variance of data in clusters). A data set for clustering consisted of the following PQ parameters:

- frequency variation (f)
- voltage variation (U)
- short-term flicker severity (P_{st})
- asymmetry (k_{u2})
- total harmonic distortion in voltage (THDu)
- active power level (P)

The indicated database consists of parameters, which are considered in the classical PQ assessment in accordance with the standard EN 50160 [81] but were extended to the active power in the measuring points. The noticeable change was the use of short-term flicker severity in place of long-term flicker severity. This change is connected with the time aggregations of the parameters; the long-term severity has 2 h, and the short-term one has 10 min [82,83]. Thus, the application of short term flicker severity enables a database consisting of parameters that are aggregated with 10 min intervals to be built, as is demanded in the standard of International Electrotechnical Commission (IEC) 61000-4-30 [80]. The analyzed measurement data were divided into flagged and unflagged data in accordance with the flagging concept of the standard IEC 61000-4-30 [80]. The data that was input to the CA were free of voltage events.

Additionally, due to the feature of the Ward algorithm that involves the fact that clustering is conducted in order to connect data concentrated in an average value until the data has similar values (range), the standardization process was proposed. The standardization of the parameters aims to obtain unified values by dividing the current value of a particular element of the time series by their maximum values. The decision concerning standardizing data to the average value reduces the problem with regards to different ranges and units of the PQ parameters. The standardize division 0–1 assures the possibility of comparing the changeability of the parameters.

3.3.2. Clustering to Indicate Different Working Conditions of the EPN

For the defined input database, the clustering with the Ward algorithm was carried out using the Statistica 13 program (StatSoft Polska, Kraków Polska). Figure 4 presents the CA dendrogram. The time results of clustering are presented in Figures 5–9, which show a defined final number of clusters equal to 2, 3, 4, 5, and 6. This selection of the number of clusters was realized using the dendrogram (Figure 4). The authors decided to indicate the cluster that has a connection distance greater than 100. Thus, no clusters equal or less than 6 were investigated. In the figures, the "virtual" cluster 0 was defined, which represents the data that was flagged in the initial stage. Using knowledge about the object, different working conditions, which may affect the data classification, were defined:

- working or non-working of distributed generation (G2, G3)—the knowledge was obtained from a monitoring system of gas-steam units:
    - working of DG: from 27.04, hour 00:00, to 08.05, hour 06:00; day 13.05, hours 11.00–12.00; day 22.05, hours 13.20–16.50
- reconfiguration of the network, the supply of main loads was relocated between substations—the knowledge was obtained from the Supervisory Control And Data Acquisition (SCADA) system:
    - from 12.05.2017, hour 16.10 to 14.05.2017, hour 22:20
- maintenance breaks that are connected to the mining industry's working schedule—checking the technical conditions of machines, a shift timetable, working on weekends:
    - each Monday–Saturday, hours 6.00–10.00: maintenance break during the first shift
    - each Saturday, hour 22.00 to Monday, hour 6.00: the weekend character of working

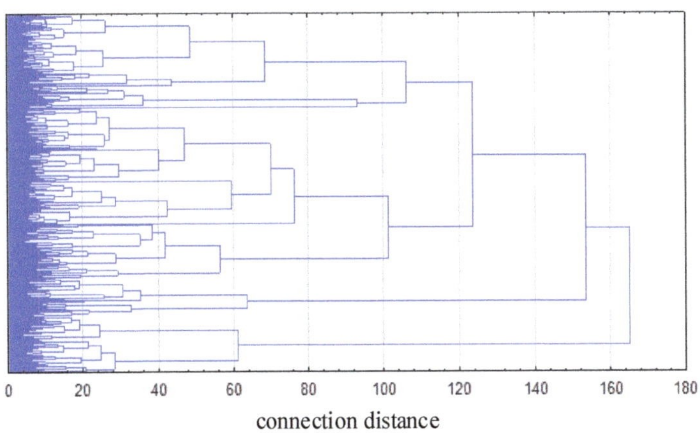

Figure 4. Dendrogram of the CA using the Ward algorithm.

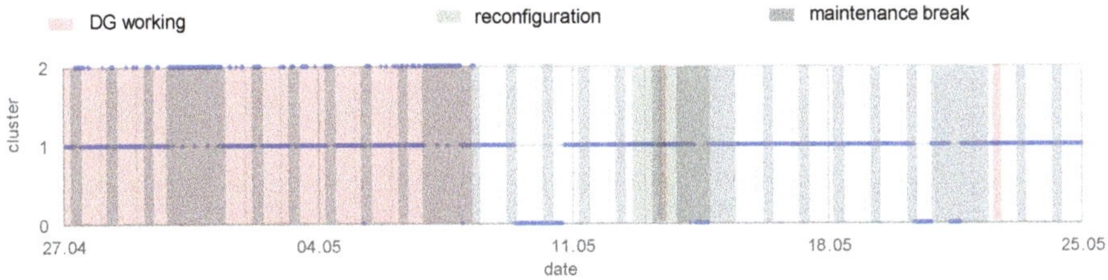

Figure 5. Cluster analysis results for the final number of clusters equal to 2.

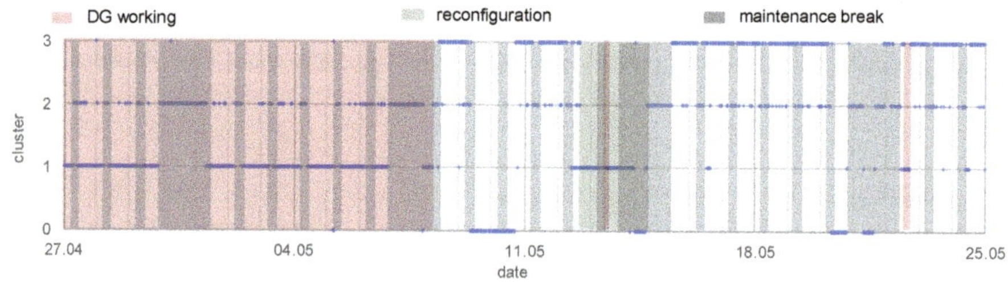

Figure 6. Cluster analysis results for the final number of clusters equal to 3.

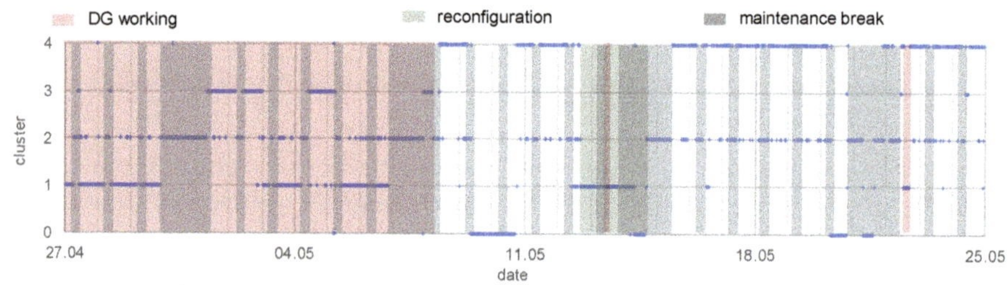

Figure 7. Cluster analysis results for the final number of clusters equal to 4.

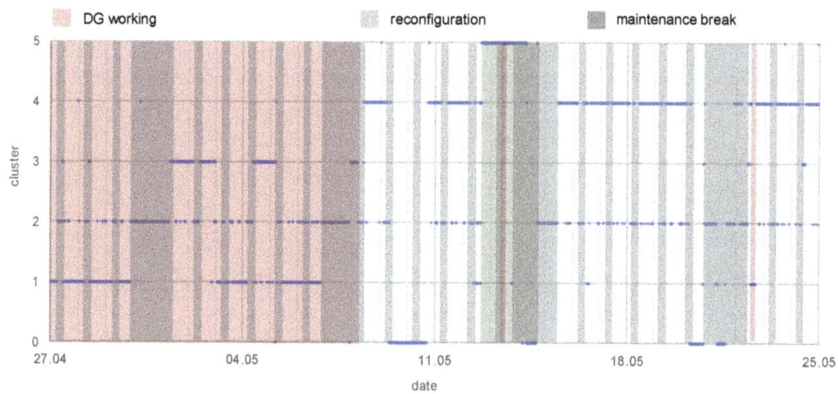

Figure 8. Cluster analysis results for the final number of clusters equal to 5.

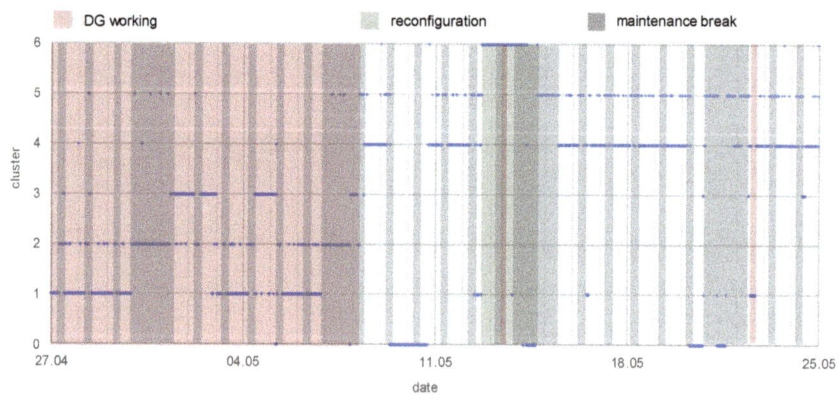

Figure 9. Cluster analysis results for the final number of clusters equal to 6.

Table 1 shows a summary of the analyzed working conditions and the assignment of clusters. The three conditions previously mentioned (DG working, reconfiguration, maintenance breaks) were indicated, and one unknown condition was observed. This unknown condition was indicated for the final number of clusters equal to at least 4. The reconfiguration of the EPN connection was indicated for the final number of clusters equal to 5. The impact of the DG and maintenance breaks was observed for all the presented classifications.

Table 1. The connection between the number of clusters and the working conditions of the electrical power network.

Condition	Final Number of Cluster				
	2	3	4	5	6
DG working	x	x	x	x	x
reconfiguration				x	x
maintenance breaks	x	x	x	x	x
other unknown condition			x	x	x

A further investigation is carried out for the final number of clusters equal to 6, although it may be realized for the other numbers too.

The analysis of the cluster assignment to the working conditions indicated that:

- c1: DG is working, exploitation time
- c2: DG is working, maintenance breaks time
- c3: DG is working, unknown working condition
- c4: DG is non-working, exploitation time
- c5: DG is non-working, maintenance breaks time
- c6: DG is non-working, reconfiguration of the network

There is an obvious question concerning which of the input parameters was important with regards to the obtained final classification. Thus, the predictor importance analysis using the Statistica 13 software (in accordance with the guidelines of a StatSoft Polska [78] and Breiman et. al. [84]) was realized for the classification of the 6 clusters. The results are presented in Figure 10. The results show that the highest impact (importance rate > 0.7) is for:

- the active power level for the transformers T1, T2, and T3
- the total harmonic distortion in the voltage for transformer T3 and the welding machine—WM
- the short-term flicker severity for the transformers T2, T3, and the welding machine—WM

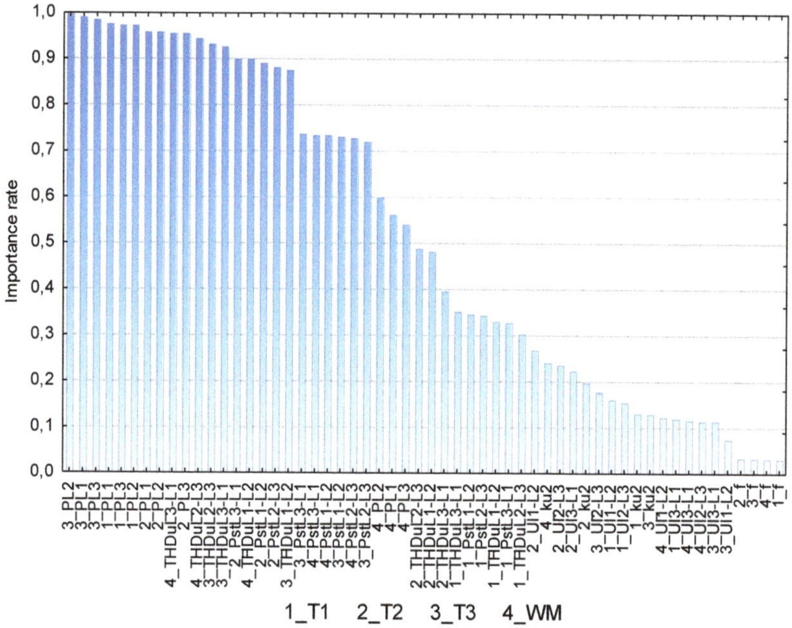

Figure 10. Importance rate of the factors to the output of the cluster analysis results for the final number of clusters equal to 6.

3.3.3. Qualitative Assessment of Clusters

A comparison of all the measurement points for each parameter in the database would lead to the analysis of the changeability of 48 parameters for each of the six clusters. Therefore, the authors suggest only analyzing those PQ parameters that were indicated as important with regards to the obtained classification (according to the predictor importance rate).

Table 2 contains the comparison of the selected PQ parameters for each cluster in terms of the mean, minimal, maximal, and standard deviation values.

Table 2. Comparison of the PQ level for different clusters.

Measurement Point	Parameter	Value	c1	c2	c3	c4	c5	c6
T2	P_{st} L1-L2	minimal	0.16	0.13	0.12	0.14	0.09	0.08
		maximal	0.50	0.30	0.54	1.42	0.74	0.57
		mean	0.22	0.21	0.18	0.24	0.26	0.11
		standard deviation	0.03	0.03	0.04	0.05	0.07	0.05
	P_{st} L2-L3	minimal	0.16	0.12	0.12	0.14	0.09	0.08
		maximal	0.50	0.29	0.53	2.07	0.73	0.38
		mean	0.22	0.21	0.18	0.24	0.26	0.11
		standard deviation	0.03	0.03	0.04	0.08	0.08	0.05
	P_{st} L3-L1	minimal	0.17	0.13	0.13	0.14	0.10	0.08
		maximal	0.50	0.30	0.52	3.54	1.06	0.53
		mean	0.23	0.21	0.19	0.25	0.27	0.11
		standard deviation	0.03	0.03	0.04	0.11	0.08	0.05
T3	P_{st} L1-L2	minimal	0.13	0.13	0.15	0.14	0.13	0.18
		maximal	0.47	0.52	0.74	0.86	0.60	0.53
		mean	0.30	0.18	0.30	0.39	0.29	0.33
		standard deviation	0.03	0.05	0.05	0.06	0.08	0.06
	P_{st} L2-L3	minimal	0.14	0.14	0.16	0.15	0.14	0.19
		maximal	0.45	0.49	0.80	2.01	0.74	0.56
		mean	0.31	0.19	0.31	0.43	0.32	0.35
		standard deviation	0.03	0.05	0.05	0.09	0.09	0.07
	P_{st} L3-L1	minimal	0.13	0.13	0.16	0.15	0.14	0.20
		maximal	0.46	0.48	0.81	1.93	1.10	0.70
		mean	0.32	0.19	0.33	0.44	0.32	0.37
		standard deviation	0.04	0.06	0.05	0.08	0.09	0.07
	THDu L1-L2	minimal	0.48	0.39	0.87	0.47	0.41	0.57
		maximal	1.20	0.99	4.99	1.50	1.11	1.38
		mean	0.67	0.56	1.53	0.83	0.64	0.80
		standard deviation	0.07	0.08	0.27	0.08	0.12	0.09
	THDu L2-L3	minimal	0.49	0.39	0.89	0.48	0.45	0.62
		maximal	1.23	0.99	5.23	1.56	1.13	1.44
		mean	0.68	0.55	1.57	0.86	0.68	0.84
		standard deviation	0.07	0.08	0.29	0.08	0.12	0.09
	THDu L3-L1	minimal	0.49	0.38	0.91	0.49	0.41	0.58
		maximal	1.28	1.02	4.87	1.63	1.18	1.50
		mean	0.70	0.55	1.63	0.89	0.67	0.87
		standard deviation	0.08	0.08	0.29	0.09	0.14	0.10
WM	P_{st} L1-L2	minimal	0.14	0.14	0.16	0.15	0.14	0.19
		maximal	0.47	0.54	0.78	6.84	0.64	0.56
		mean	0.31	0.19	0.32	0.43	0.31	0.35
		standard deviation	0.03	0.05	0.05	0.20	0.08	0.07
	P_{st} L2-L3	minimal	0.13	0.13	0.16	0.15	0.14	0.19
		maximal	0.45	0.49	0.79	6.89	0.72	0.57
		mean	0.31	0.19	0.32	0.43	0.31	0.35
		standard deviation	0.03	0.06	0.05	0.20	0.08	0.07
	P_{st} L3-L1	minimal	0.14	0.13	0.16	0.15	0.14	0.19
		maximal	0.46	0.46	0.78	6.84	1.12	0.66
		mean	0.31	0.18	0.31	0.43	0.31	0.35
		standard deviation	0.03	0.05	0.05	0.20	0.09	0.07

Table 2. *Cont.*

	THDu L1-L2	minimal	0.46	0.36	0.55	0.48	0.40	0.56
		maximal	1.23	0.99	2.40	1.54	1.16	1.42
		mean	0.67	0.53	1.56	0.85	0.65	0.81
		standard deviation	0.08	0.08	0.22	0.08	0.13	0.10
	THDu L2-L3	minimal	0.45	0.36	0.59	0.49	0.43	0.59
		maximal	1.23	0.96	2.40	1.55	1.13	1.44
		mean	0.65	0.52	1.54	0.85	0.67	0.84
		standard deviation	0.08	0.08	0.23	0.08	0.13	0.10
	THDu L3-L1	minimal	0.45	0.36	0.58	0.49	0.42	0.59
		maximal	1.22	0.94	2.37	1.53	1.12	1.42
		mean	0.65	0.52	1.50	0.86	0.67	0.85
		standard deviation	0.08	0.08	0.22	0.08	0.13	0.10

where:

minimal—the minimal value of the parameter that may be found for the observed cluster
maximal—the maximal value of the parameter that may be found for the observed cluster
mean—the mean value calculated from all the data for the observed cluster
standard deviation—the standard deviation calculated from all the data for the observed cluster.

A comparison of the level of the PQ parameters for different clusters is equivalent to the comparison of the different working conditions of an electrical power network. The examples of such a comparison may be as follows:

- (c1 with c2) and (c4 with c5)—> comparison of time with the different characters of the company that iss working (exploitation vs. maintenance break). It could be observed that the mean value of P_{st} for T3 and WM is lower during the maintenance break. Therefore, in terms of flicker severity, the time of maintenance is better.
- (c1 with c2) and (c4 with c5)—>comparison of time with the different characters of the company that is working (exploitation vs. maintenance break). It could be observed that the mean value of THDu for T3 and WM are lower during the maintenance break. Therefore, in terms of the harmonic content, the time of maintenance is better.
- (c1 with c4) and (c2 with c5)—> comparison of time with the different characters of the working DG. It could be observed that P_{st} for T3 and WM is lower for the time when the DG is working (c1, c2) compared to when the DG is switched off (c4, c5). Therefore, in terms of flicker severity, the time when the DG is working is better.
- c3 with all other clusters—> this unknown working condition represents the time when the THDu level for T3 and WM is higher than for the other clusters.
- c6 with all other clusters—> the reconfiguration that represents the time when P_{st} for T2 is very low. This is in agreement with the fact that T2 was underloaded, and therefore, the flicker is small

The presented examples about the comparison of the level of the PQ parameters for different clusters assure simplified information concerning the differences between working conditions. However, the working condition for defining the cluster c3 is unknown, but due to the indicated analysis, it is possible to define that during this time there was a higher than normal level of harmonics for T3 and WM. Thanks to this, attention could be paid to this time in order to find the reason for such high harmonic content and to reduce it in the future. Additionally, after automatic classification of the data, it is possible to show the impact of DG on the level of power quality in the electrical power network of the mining industry.

3.3.4. Reduction of the Input Database Size—Case Study

The natural question is, "is it possible to reduce or change the structure of the input database without losing the most important information". The first idea is just to exclude some parameters. However, the proposed, complete database includes, all-important points of the classical PQ parameters. Thus, excluding any of them would not seem to be adequate from the technical point of view.

In this research, the objects are represented by similar phase-to-phase values. Thus, the analysis of only one "new-multiphase" value was conducted. Moreover, the way of conducting this may be different. The minimal, maximal, mean, or median value from three phase-to-phase values may be selected. However, in this research, the authors decided to use the mean value. Thus, for each 10 min data of:

- voltage
- short-term flicker severity
- total harmonic distortion in voltage
- active power

the mean value from all three phase-to-phase values was calculated.

After such a reduction—from 16 input parameters (complete database) for each measurement point to six input parameters (reduced database)—clustering was conducted. The result of the obtained cluster using the six-parameter database, in comparison to 14-parameter clustering, is presented in Table 3. Generally, the results of this reduction in terms of indicating the same working condition for more than two clusters are positive. The obtained classification has the same result for at least 94.9% of data. The only negative classification was obtained for two clusters. The averaged data during the division to two clusters was not sensitive for DG impact.

Table 3. Comparison of clustering results for the completed database to the reduced one.

Final Number of Clusters	Do Results Indicate the Same Working Conditions?	Percent of the Data Assigned to the Same Cluster
2	no *	–
3	yes	95.7
4	yes	95.1
5	yes	95.0
6	yes	94.9

* no impact of DG is observable, only the maintenance is noticeable.

Additionally, the predictor importance for six clusters was defined. Figure 11 presents the importance rate for both classifications—(a) reduced input database, (b) complete input database. Generally, regarding the 0.7 importance rate level (noticeable importance rate), the same parameters were indicated:

- transformer T1—active power
- transformer T2—active power
- transformer T3—active power, total harmonic distortion in voltage, short-term flicker severity
- welding machine WM—active power, total harmonic distortion in voltage, short-term flicker severity

The only excluded parameter is the short-term flicker severity for transformer T2. However, the importance rate is close to 0.7.

To summarize, the size of the database has been reduced from 14 parameters to six parameters, and the obtained results are generally similar.

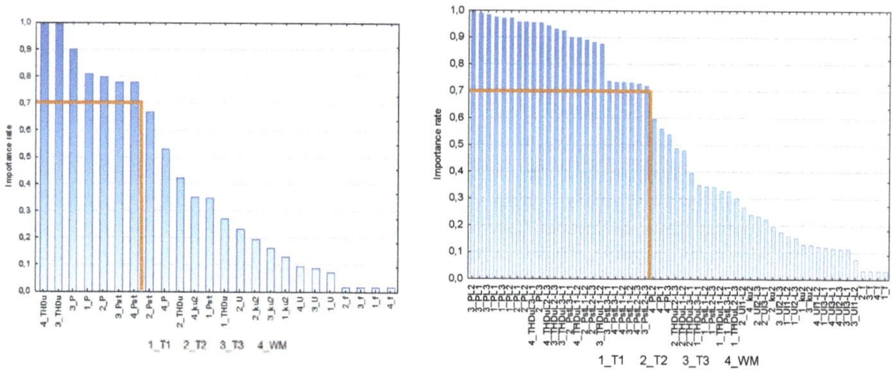

(a) (b)

Figure 11. Importance rate for six clusters for (**a**) reduced input database; (**b**) complete input database.

3.4. Discussion

The data mining technique presented in the article is the cluster analysis. The Ward algorithm was selected as an example of the hierarchical approach. During the data preparation stage, it was necessary to uniform the data aggregation time (selection of P_{lt} to P_{st}), as well as to standardize the parameter values. For the prepared data set containing both the PQ parameters and the active power level, cluster analysis was conducted.

As a result of the cluster analysis, a dendrogram was obtained, which was illegible for the initial stages of agglomeration due to a large amount of input data. This is an unquestionable disadvantage of the hierarchical approach, but it is worth noting that it provides a division of data regardless of the final number of the obtained clusters. Additionally, on the dendrogram, there is a simple possibility of selecting the final number of clusters using methods indicated in the literature, e.g., Aggarwal [79].

Another important element of the article was to indicate the conditions that influenced the data division. On the basis of knowledge about the object, the conditions of distributed generation working, reconfiguration, and maintenance breaks were known. However, the obtained classification indicated that, in terms of the PQ level, the relevant condition was not known. It is worth highlighting the fact that the Ward algorithm is sensitive to the impact of the distributed generation on the technical conditions of the electrical power network, which confirms that the research aim was specified correctly.

The next element of the article was the analyses of the parameters that have a higher impact on the data classification. The obtained results indicated the importance of an active power level, as well as the harmonic level and flicker. The voltage variations, voltage, and frequency levels had a small impact on the classification.

Then, after obtaining the importance ranking, a comparison of the clusters in terms of the selected PQ parameters was carried out. The obtained results presented the impact of DG on the EPN. The impact of DG was indicated as positive regarding PQ. The unknown working condition was described as a time with high total harmonic distortion at the voltage level. Thus, the analysis of only this selected period of time may help to decrease the problem with harmonic pollution.

The last part of the research concerned the possibility of reducing the input database without losing the information obtained from the clustering. The authors proposed reducing the three phase-to-phase values to one mean value. Then, the comparison of the reduced input database to the completed one was conducted. The obtained classifications were similar. Around 95% of data was connected to the same clusters for both input databases and classification to more than two groups. The presented approach decreased the size of the input database by 57% (from fourteen to six parameters) without losing any data features.

The presented in-article object represents a symmetrical network, although, the method may be realized successfully for highly asymmetrical grids. Thus, if any of the phase-to-phase value was changed, the mean value of all parameters also changed. The CA is sensitive for the differences so this situation would also be indicated. The only disadvantage of this method is that there would be no information on which phase caused this situation, thus the analysis of raw data, but for the indicated period of time, is desirable.

4. Conclusions

The article presents the application of cluster analysis to long-term power quality measurements obtained in an electrical power network of the mining industry with distributed generation. The selected algorithm, due to its sensitivity to data dissimilarity, was the Ward algorithm. The article contains a discussion of the pros and cons of the hierarchical approach.

The article also contains the analysis of the sensitivity of different (known) working conditions of an electrical power network of the mining industry to the obtained classification. Conditions such as the impact of distributed generation, reconfiguration appearance, or the character of the object schedule (exploitation or maintenance breaks) are indicated. Additionally, the ranking of the impact of the parameter on the classification was conducted using predictor analysis. This analysis indicated that the level of active power, harmonic pollution, and flicker are important with regards to the obtained classification.

The obtained classification indicated the unknown working condition. After the comparison with other groups, the unknown condition was indicated as a high harmonic pollution period of time. Thanks to this, it is possible to analyze a short period of time to find the problem with harmonic pollution in an electrical power network of the mining industry.

The article contains the proposition of reducing a database concerning the calculation of one value that represents three phase-to-phase values. The results were similar (close to 95%), and the calculations were reduced by over 57%.

The presented approach of obtaining automatic data classification with regards to different working conditions (especially distributed generation or the harmonic pollution problem) is an important element of a smart grid. It is worth noting that the presented approach is conducted for area-related analysis—four different measuring points that are considered as common input data.

Author Contributions: Conceptualization, M.J. and T.S.; methodology, M.J. and T.S.; software, M.J.; validation, M.J., T.S., K.B.; formal analysis, M.J. and E.J.; investigation, M.J.; resources, M.J., T.S., K.B.; data curation, M.J.; writing—original draft preparation, M.J.; writing—review and editing, T.S.; visualization, M.J. and E.J.; supervision, T.S., Z.L.; project administration, T.S.; funding acquisition, Z.L. All authors have read and agreed to the published version of the manuscript.

Funding: This research was funded by the Chair of Electrical Engineering Fundamentals (K38W05D02), Wroclaw University of Technology, Wroclaw, Poland.

Acknowledgments: The authors would like to thank KGHM Polska Miedź S.A. for support.

Conflicts of Interest: The authors declare no conflict of interest.

References

1. Oncioiu, I.; Căpuşneanu, S.; Türkeș, M.; Topor, D.; Constantin, D.-M.; Marin-Pantelescu, A.; Ștefan Hint, M. The Sustainability of Romanian SMEs and Their Involvement in the Circular Economy. *Sustainability* **2018**, *10*, 2761. [CrossRef]
2. Türkeș, M.; Oncioiu, I.; Aslam, H.; Marin-Pantelescu, A.; Topor, D.; Căpușneanu, S. Drivers and Barriers in Using Industry 4.0: A Perspective of SMEs in Romania. *Processes* **2019**, *7*, 153. [CrossRef]
3. Oncioiu, I.; Bunget, O.C.; Türkeș, M.C.; Căpușneanu, S.; Topor, D.I.; Tamaș, A.S.; Rakos, I.-S.; Hint, M.Ș. The Impact of Big Data Analytics on Company Performance in Supply Chain Management. *Sustainability* **2019**, *11*, 4864. [CrossRef]
4. Salkuti, S.R. A survey of big data and machine learning. *Int. J. Electr. Comput. Eng.* **2020**, *10*, 575. [CrossRef]

5. Ghorbanian, M.; Dolatabadi, S.H.; Siano, P. Big Data Issues in Smart Grids: A Survey. *IEEE Syst. J.* **2019**, *13*, 4158–4168. [CrossRef]
6. Dhupia, B.; Usha Rani, M.; Alameen, A. The role of big data analytics in smart grid management. In Proceedings of the 2nd International Conference on Computing, Communications Data Engineering CCODE 2019, Tirupati, India, 1–2 February 2019; Volume 1054, pp. 403–412.
7. Ding, Y. Analysis of Operation and Maintenance of Power Distribution Network Management Technology Under the Background of Big Data Era. In *International Conference on Big Data Analytics for Cyber-Physical-Systems*; Springer: Singapore, 2020; pp. 610–615.
8. Jasiński, M.; Sikorski, T.; Borkowski, K. Clustering as a tool to support the assessment of power quality in electrical power networks with distributed generation in the mining industry. *Electr. Power Syst. Res.* **2019**, *166*, 52–60. [CrossRef]
9. Jasiński, M.; Sikorski, T.; Kostyła, P.; Leonowicz, Z.; Borkowski, K. Combined Cluster Analysis and Global Power Quality Indices for the Qualitative Assessment of the Time-Varying Condition of Power Quality in an Electrical Power Network with Distributed Generation. *Energies* **2020**, *13*, 2050. [CrossRef]
10. Strack, J.L.; Carugati, I.; Orallo, C.M.; Maestri, S.O.; Donato, P.G.; Funes, M.A. Three-phase voltage events classification algorithm based on an adaptive threshold. *Electr. Power Syst. Res.* **2019**, *172*, 167–176. [CrossRef]
11. Shikhin, V.A.; Kochengin, A.E.; Pavliuk, G.P. Significant Events Detection and Identification through Electrical Grid Load Profile. In Proceedings of the 2018 IEEE Renewable Energies, Power Systems & Green Inclusive Economy (REPS-GIE), Casablanca, Morocco, 23–24 April 2018; pp. 1–5.
12. Ucar, F.; Alcin, O.F.; Dandil, B.; Ata, F. Power quality event detection using a fast extreme learning machine. *Energies* **2018**, *11*, 145. [CrossRef]
13. Biswal, B.; Biswal, M.; Mishra, S.; Jalaja, R. Automatic classification of power quality events using balanced neural tree. *IEEE Trans. Ind. Electron.* **2014**, *61*, 521–530. [CrossRef]
14. Jasiński, M.; Sikorski, T.; Borkowski, K. Application of cluster analysis to identification flagged power quality measurements in area-related approach. Zastosowanie eksploracji danych do identyfikacji oznaczonych wyników pomiaru jakosci energii elektrycznej w ujeciu obszarowym. *Prz. Elektrotechniczny* **2020**, *3*, 9–12.
15. Balouji, E.; Salor, O. Classification of power quality events using deep learning on event images. In Proceedings of the 3rd International Conference on Pattern Analysis Image Analysis IPRIA 2017, Shahrekord, Iran, 19–20 April 2017.
16. Dangar, B.; Josh, S.K. Interpretation of Urban Power Consumers Behaviors to Predict Power Loss in Summer. *Int. J. Eng. Adv. Technol.* **2019**, *9*, 563–565.
17. Yun, Z.; Mengting, Y.; Junjie, L.; Ji, C.; Penghui, H. Line loss calculation of low-voltage districts based on improved K-Means. In Proceedings of the 2018 IEEE International Conference on Power System Technology (POWERCON), Beijing, China, 24–26 October 2018; pp. 4578–4583.
18. Yao, M.; Zhu, Y.; Li, J.; Wei, H.; He, P. Research on Predicting Line Loss Rate in Low Voltage Distribution Network Based on Gradient Boosting Decision Tree. *Energies* **2019**, *12*, 2522. [CrossRef]
19. Menezes, A.G.C.; Almeida, O.M.; Barbosa, F.R. Use of decision tree algorithms to diagnose incipient faults in power transformers. In Proceedings of the 2018 IEEE Simposio Brasileiro de Sistemas Eletricos (SBSE), Niteroi, Brazil, 12–16 May 2018; pp. 1–6.
20. Liu, C.H.; Chen, T.L.; Yao, L.T.; Wang, S.Y. Using data mining to dissolved gas analysis for power transformer fault diagnosis. In Proceedings of the 2012 IEEE International Conference on Machine Learning and Cybernetics, Xian, China, 15–17 July 2012; pp. 1952–1957.
21. Basuki, A. Suwarno Online Dissolved Gas Analysis of Power Transformers Based on Decision Tree Model. In Proceedings of the 2018 IEEE Conference on Power Engineering and Renewable Energy (ICPERE), Solo, Indonesia, 29–31 October 2018; pp. 1–6.
22. Ren, F.; Si, S.; Cai, Z.; Zhang, S. Transformer fault analysis based on Bayesian networks and importance measures. *J. Shanghai Jiaotong Univ.* **2015**, *20*, 353–357. [CrossRef]
23. Cheng, L.; Yu, T. Dissolved Gas Analysis Principle-Based Intelligent Approaches to Fault Diagnosis and Decision Making for Large Oil-Immersed Power Transformers: A Survey. *Energies* **2018**, *11*, 913. [CrossRef]
24. Almeida, V.A.; Pessanha, J.F.M.; Caloba, L.P. Load data cleaning with data mining techniques. In Proceedings of the 2018 IEEE Simposio Brasileiro de Sistemas Eletricos (SBSE), Niteroi, Brazil, 12–16 May 2018; pp. 1–6.

25. Kotriwala, A.M.; Hernandez-Leal, P.; Kaisers, M. Load Classification and Forecasting for Temporary Power Installations. In Proceedings of the 2018 IEEE PES Innovative Smart Grid Technologies Conference Europe (ISGT-Europe), Sarajevo, Bosnia, 21–25 October 2018; pp. 1–6.
26. Cerne, G.; Dovzan, D.; Skrjanc, I. Short-Term Load Forecasting by Separating Daily Profiles and Using a Single Fuzzy Model Across the Entire Domain. *IEEE Trans. Ind. Electron.* **2018**, *65*, 7406–7415. [CrossRef]
27. Lei, J.; Jin, T.; Hao, J.; Li, F. Short-term load forecasting with clustering–regression model in distributed cluster. *Clust. Comput.* **2019**, *22*, 10163–10173. [CrossRef]
28. Fahiman, F.; Erfani, S.M.; Leckie, C. Robust and Accurate Short-Term Load Forecasting: A Cluster Oriented Ensemble Learning Approach. In Proceedings of the 2019 IEEE International Joint Conference on Neural Networks (IJCNN), Budapest, Hungary, 14–19 July 2019; pp. 1–8.
29. Arun Jees, S.; Gomathi, V. Load forecasting for smart grid using non-linear model in Hadoop distributed file system. *Clust. Comput.* **2019**, *22*, 13533–13545. [CrossRef]
30. Rajabi, A.; Eskandari, M.; Ghadi, M.J.; Li, L.; Zhang, J.; Siano, P. A comparative study of clustering techniques for electrical load pattern segmentation. *Renew. Sustain. Energy Rev.* **2019**, *120*, 109628. [CrossRef]
31. Verdu, S.V.; Garcia, M.O.; Senabre, C.; Marin, A.G.; Franco, F.J.G. Classification, Filtering, and Identification of Electrical Customer Load Patterns Through the Use of Self-Organizing Maps. *IEEE Trans. Power Syst.* **2006**, *21*, 1672–1682. [CrossRef]
32. Le Ray, G.; Pinson, P. Online adaptive clustering algorithm for load profiling. *Sustain. Energy Grids Netw.* **2019**, *17*, 100181. [CrossRef]
33. Chicco, G. Overview and performance assessment of the clustering methods for electrical load pattern grouping. *Energy* **2012**, *42*, 68–80. [CrossRef]
34. Ramdasi, A.P.; Mehata, K.M. Improved Text Mining Algorithm for Fault Detection using Combined D-Matrix. *Int. J. Recent Technol. Eng.* **2019**, *8*, 1376–1379.
35. Gao, T.; Boguslawski, B.; Marié, S.; Béguery, P.; Thebault, S.; Lecoeuche, S. Data mining and data-driven modelling for Air Handling Unit fault detection. In *E3S Web of Conferences*; EDP Sciences: Jules, France, 2019; Volume 111.
36. Chen, L.; Xu, G.; Zhang, Q.; Zhang, X. Learning deep representation of imbalanced SCADA data for fault detection of wind turbines. *Measurement* **2019**, *139*, 370–379. [CrossRef]
37. Ranjbar, S.; Jamali, S. Fault detection in microgrids using combined classification algorithms and feature selection methods. In Proceedings of the 13th International Conference on Protection and Automation of Power System, IPAPS 2019, Tehran, Iran, 31 December 2019–1 January 2020; Institute of Electrical and Electronics Engineers Inc., School of Electrical Engineering, Iran University of Science and Technology (IUST): Tehran, Iran, 2019; pp. 17–21.
38. Silva, S.; Costa, P.; Gouvea, M.; Lacerda, A.; Alves, F.; Leite, D. High impedance fault detection in power distribution systems using wavelet transform and evolving neural network. *Electr. Power Syst. Res.* **2018**, *154*, 474–483. [CrossRef]
39. Sun, C.; Wang, X.; Zheng, Y.; Zhang, F. A framework for dynamic prediction of reliability weaknesses in power transmission systems based on imbalanced data. *Int. J. Electr. Power Energy Syst.* **2020**, *117*, 105718. [CrossRef]
40. Sun, C.; Wang, X.; Zheng, Y. Data-driven approach for spatiotemporal distribution prediction of fault events in power transmission systems. *Int. J. Electr. Power Energy Syst.* **2019**, *113*, 726–738. [CrossRef]
41. Pal, A.; Kumar, M. DLME: Distributed Log Mining Using Ensemble Learning for Fault Prediction. *IEEE Syst. J.* **2019**, *13*, 3639–3650. [CrossRef]
42. Cynthia, S.T.; Ripon, S.H. Predicting and Classifying Software Faults. In Proceedings of the 2019 7th International Conference on Computer and Communications Management—ICCCM 2019, Bangkok, Thailand, 27–29 July 2019; ACM Press: New York, NY, USA, 2019; pp. 143–147.
43. Rathod, R.R.; Garg, R.D. Regional electricity consumption analysis for consumers using data mining techniques and consumer meter reading data. *Int. J. Electr. Power Energy Syst.* **2016**, *78*, 368–374. [CrossRef]
44. Benítez, I.; Quijano, A.; Díez, J.-L.; Delgado, I. Dynamic clustering segmentation applied to load profiles of energy consumption from Spanish customers. *Int. J. Electr. Power Energy Syst.* **2014**, *55*, 437–448. [CrossRef]
45. Cil, I. Consumption universes based supermarket layout through association rule mining and multidimensional scaling. *Expert Syst. Appl.* **2012**, *39*, 8611–8625. [CrossRef]

46. Zhang, G.; Wang, G.G.; Farhangi, H.; Palizban, A. Data mining of smart meters for load category based disaggregation of residential power consumption. *Sustain. Energy Grids Netw.* **2017**, *10*, 92–103. [CrossRef]
47. Jain, P.K.; Quamer, W.; Pamula, R. *Electricity Consumption Forecasting Using Time Series Analysis BT—Advances in Computing and Data Sciences*; Singh, M., Gupta, P.K., Tyagi, V., Flusser, J., Ören, T., Eds.; Springer: Singapore, 2018; pp. 327–335.
48. Yildiz, B.; Bilbao, J.I.; Dore, J.; Sproul, A.B. Recent advances in the analysis of residential electricity consumption and applications of smart meter data. *Appl. Energy* **2017**, *208*, 402–427. [CrossRef]
49. Sheng, H.; Xiao, J.; Cheng, Y.; Ni, Q.; Wang, S. Short-Term Solar Power Forecasting Based on Weighted Gaussian Process Regression. *IEEE Trans. Ind. Electron.* **2018**, *65*, 300–308. [CrossRef]
50. Anderson, W.W.; Yakimenko, O.A. Using neural networks to model and forecast solar PV power generation at Isle of Eigg. In Proceedings of the 2018 IEEE 12th International Conference on Compatibility, Power Electronics and Power Engineering (CPE-POWERENG 2018), Doha, Qatar, 10–12 April 2018; pp. 1–8.
51. Yao, S.; Pan, L.; Yu, Z.; Kang, Q.; Zhou, M. Hierarchically Non-continuous Regression Prediction for Short-Term Photovoltaic Power Output. In Proceedings of the 2019 IEEE 16th International Conference on Networking, Sensing and Control (ICNSC), Banff, AB, Canada, 9–11 May 2019; pp. 379–384.
52. Monfared, M.; Fazeli, M.; Lewis, R.; Searle, J. Fuzzy Predictor with Additive Learning for Very Short-Term PV Power Generation. *IEEE Access* **2019**, *7*, 91183–91192. [CrossRef]
53. Su, C.; Hu, Z. Reliability assessment for Chinese domestic wind turbines based on data mining techniques. *Wind Energy* **2018**, *21*, 198–209. [CrossRef]
54. Aikhuele, D.O. Intuitionistic fuzzy model for reliability management in wind turbine system. *Appl. Comput. Inform.* **2018**. [CrossRef]
55. Uma, J.; Muniraj, C.; Sathya, N. Diagnosis of Photovoltaic (PV) Panel Defects Based on Testing and Evaluation of Thermal Image. *J. Test. Eval.* **2019**, *47*, 4249–4262. [CrossRef]
56. Harrou, F.; Dairi, A.; Taghezouit, B.; Sun, Y. An unsupervised monitoring procedure for detecting anomalies in photovoltaic systems using a one-class Support Vector Machine. *Sol. Energy* **2019**, *179*, 48–58. [CrossRef]
57. Du, S.; Li, M.; Han, S.; Shi, J.; Li, H. Multi-Pattern Data Mining and Recognition of Primary Electric Appliances from Single Non-Intrusive Load Monitoring Data. *Energies* **2019**, *12*, 992. [CrossRef]
58. Parvizimosaed, M.; Farmani, F.; Rahimi-Kian, A.; Monsef, H. A multi-objective optimization for energy management in a renewable micro-grid system: A data mining approach. *J. Renew. Sustain. Energy* **2014**, *6*. [CrossRef]
59. Ai, S.; Chakravorty, A.; Rong, C. Household Power Demand Prediction Using Evolutionary Ensemble Neural Network Pool with Multiple Network Structures. *Sensors* **2019**, *19*, 721. [CrossRef] [PubMed]
60. Singh, S.; Yassine, A. Mining Energy Consumption Behavior Patterns for Households in Smart Grid. *IEEE Trans. Emerg. Top. Comput.* **2019**, *7*, 404–419. [CrossRef]
61. El Mrabet, Z.; El Ghazi, H.; Kaabouch, N. A Performance Comparison of Data Mining Algorithms Based Intrusion Detection System for Smart Grid. In Proceedings of the 2019 IEEE International Conference on Electro Information Technology (EIT), Brookings, SD, USA, 20–22 May 2019; pp. 298–303.
62. Gupta, S.; Sabitha, A.S.; Punhani, R. Cyber Security Threat Intelligence using Data Mining Techniques and Artificial Intelligence. *Int. J. Recent Technol. Eng.* **2019**, *8*, 6133–6140.
63. Zuo, X.; Chen, Z.; Dong, L.; Chang, J.; Hou, B. Power information network intrusion detection based on data mining algorithm. *J. Supercomput.* **2019**. [CrossRef]
64. Ahmad, T.; Chen, H.; Wang, J.; Guo, Y. Review of various modeling techniques for the detection of electricity theft in smart grid environment. *Renew. Sustain. Energy Rev.* **2018**, *82*, 2916–2933. [CrossRef]
65. Razavi, R.; Gharipour, A.; Fleury, M.; Akpan, I.J. A practical feature-engineering framework for electricity theft detection in smart grids. *Appl. Energy* **2019**, *238*, 481–494. [CrossRef]
66. Maamar, A.; Benahmed, K. Machine learning Techniques for Energy Theft Detection in AMI. In Proceedings of the 2018 International Conference on Software Engineering and Information Management—ICSIM2018, Casablanca, Morocco, 4–6 January 2018; ACM Press: New York, NY, USA, 2018; pp. 57–62.
67. Jindal, A.; Dua, A.; Kaur, K.; Singh, M.; Kumar, N.; Mishra, S. Decision Tree and SVM-Based Data Analytics for Theft Detection in Smart Grid. *IEEE Trans. Ind. Inform.* **2016**, *12*, 1005–1016. [CrossRef]
68. Han, J.; Kamber, M. *Data Mining: Concepts and Techniques*; Elsevier: Amsterdam, The Netherlands, 2011; Volume 12, ISBN 978-3-642-19720-8.

69. Larose, D. *Discovering Knowledge in Data. An Introduction to Data Mining*; John Wiley & Sons: Hoboken, NJ, USA, 2005; pp. 1–35. ISBN 9786468600.
70. Kantardzic, M. *Data Mining: Concepts, Models, Methods, and Algorithms*, 2nd ed.; John Wiley & Sons: Hoboken, NJ, USA, 2011; ISBN 9780470890455.
71. CIGRE. *Broshure 292: Data Mining Techniques and Applications in the Power Transmission Field*; CIGRE: Paris, France, 2006.
72. Olson, D.L.; Delen, D. *Advanced Data Mining Techniques*; Springer: Berlin/Heidelberg, Germany, 2008; ISBN 978-3-540-76916-3.
73. Wierzchoń, S.; Kłopotek, M. *Algorithms of Cluster Analysis*; Institute of Computer Science Polish Academy of Sciences: Warszaw, Poland, 2015; Volume 3, ISBN 9789638759627.
74. Witten, I.H.; Frank, E. *Data Mining: Practical Machine Learning Tools and Techniques*; Morgan Kaufmann Publishers: San Francisco, CA, USA, 2011; ISBN 0080890369.
75. Wu, X.; Kumar, V.; Ross, Q.J.; Ghosh, J.; Yang, Q.; Motoda, H.; McLachlan, G.J.; Ng, A.; Liu, B.; Yu, P.S.; et al. Top 10 algorithms in data mining. *Knowl. Inf. Syst.* **2008**, *14*, 1–37. [CrossRef]
76. Sneath, P.H.; Sokal, R.R. *Numerical Texonomy*; Freeman: Lanzhou, China, 1973; ISBN 9780716706977.
77. Jasiński, M.; Borkowski, K.; Sikorski, T.; Kostyla, P. Cluster Analysis for Long-Term Power Quality Data in Mining Electrical Power Network. In Proceedings of the 2018 IEEE Progress in Applied Electrical Engineering (PAEE), Koscielisko, Poland, 18–22 June 2018; pp. 1–5.
78. Statsoft Polska StatSoft Electronic Statistic Textbook. Available online: http.:/www.statsoft.pl/textbook/stathome.html (accessed on 15 February 2020).
79. Aggarwal, C.C. *Data Mining*; Springer: Cham, Switzerland, 2015; ISBN 978-3-319-14141-1.
80. *International Electrotechnical Commission, IEC 61000 4-30 Electromagnetic Compatibility (EMC)—Part 4-30: Testing and Measurement Techniques—Power Quality Measurement Methods*; International Electrotechnical Commission: Geneva, Switzerland, 2015.
81. *British Standards Institution, EN 50160: Voltage Characteristics of Electricity Supplied by Public Distribution Network*; British Standards Institution: UK, 2010.
82. Jasiński, M.; Sikorski, T.; Kostyła, P.; Kaczorowska, D.; Leonowicz, Z.; Rezmer, J.; Szymańda, J.; Janik, P.; Bejmert, D.; Rybiański, M.; et al. Influence of Measurement Aggregation Algorithms on Power Quality Assessment and Correlation Analysis in Electrical Power Network with PV Power Plant. *Energies* **2019**, *12*, 3547. [CrossRef]
83. Jasiński, M.; Rezmer, J.; Sikorski, T.; Szymańda, J. Integration Monitoring of On-grid Photovoltaic System: Case Study. *Period. Polytech. Electr. Eng. Comput. Sci.* **2019**, *63*, 99–105. [CrossRef]
84. Breiman, L.; Friedman, J.; Stone, C.J.; Olshen, R.A. *Classification and Regression Trees*; CRC Press: Boca Raton, FL, USA, 1984.

energies

MDPI

Article

Analysis of the Power Supply Restoration Time after Failures in Power Transmission Lines

Alexander Vinogradov [1], Vadim Bolshev [1,*], Alina Vinogradova [1], Michał Jasiński [2,*], Tomasz Sikorski [2], Zbigniew Leonowicz [2], Radomir Goňo [3] and Elżbieta Jasińska [4]

1 Laboratory of Power Supply and Heat Supply, Federal Scientific Agroengineering Center VIM, 109428 Moscow, Russia; schkolamolen@gmail.com (A.V.); alinawin@rambler.ru (A.V.)
2 Department of Electrical Engineering Fundamentals, Faculty of Electrical Engineering, Wroclaw University of Science and Technology, 50-370 Wroclaw, Poland; tomasz.sikorski@pwr.edu.pl (T.S.); zbigniew.leonowicz@pwr.edu.pl (Z.L.)
3 Department of Electrical Power Engineering, Faculty of Electrical Engineering and Computer Science, VSB-Technical University of Ostrava, 708 00 Ostrava, Czech Republic; radomir.gono@vsb.cz
4 Faculty of Law, Administration and Economics, University of Wroclaw, 50-145 Wroclaw, Poland; elzbieta.jasinska@uwr.edu.pl
* Correspondence: vadimbolshev@gmail.com (V.B.); michal.jasinski@pwr.edu.pl (M.J.); Tel.: +7-499-174-85-95 (V.B.); +48-713-202-022 (M.J.)

Received: 24 April 2020; Accepted: 27 May 2020; Published: 29 May 2020

Abstract: This paper presents the analysis of power supply restoration time after failures occurring in power lines. It found that the power supply restoration time depends on several constituents, such as the time for obtaining information on failures, the time for information recognition, the time to repair failures, and the time for connection harmonization. All these constituents have been considered more specifically. The main constituents' results values of the power supply restoration time were analyzed for the electrical networks of regional power supply company "Oreolenergo", a branch of Interregional Distribution Grid Company (IDGC) of Center. The Delphi method was used for determining the time for obtaining information on failures as well as the time for information recognition. The method of mathematical statistics was used to determine the repair time. The determined power supply restoration time (5.28 h) is similar to statistical values of the examined power supply company (the deviation was equal to 9.9%). The technical means of electrical network automation capable of the reduction of the power supply restoration time have also been found. These means were classified according to the time intervals they shorten.

Keywords: power supply restoration; power supply outages; failures; time intervals; obtaining information; information recognition; connection harmonization

1. Introduction

Improving power supply efficiency from private homes to large industrial enterprises is an urgent and difficult task for power supply (PS) enterprises. This is because power supply companies often encounter problems such as the remoteness of energy consumers from power distribution points, insufficient capacity margin, depreciation of power supply equipment, and the lack of specialists involved in servicing this equipment. As a result, it leads to the increase in the number of equipment failures and the increase in PS interruptions. In turn, this translates into losses for power supply companies due to the elimination of the failure's consequences and for consumers due to a disruption of the technological process caused by the power supply outages.

This article deals with the analysis of power supply restoration after power line failures. Thus, the organization of the article is as follows. Section 2 is a literature review of the state of the art with the motivation and contribution of this paper. Section 3 introduces the problem of power supply reliability. It indicates the elements that have an impact on the total time of restoration. In Section 4, the calculation of each element is performed based on real data obtained from a Russian power supply company. Section 5 contains a discussion of the results and propositions for the technical means of automation of electrical networks to reduce the power supply restoration time. Section 6 is the conclusion part.

2. State-of-the-Art Power Supply Restoration Issues

The present research described in the literature concerns the problem of power supply reliability in different areas. The aim of this research is a power supply restoration analysis [1]. Thus, current research trends in the literature are analyzed with regards to restoration issues. The presented literature review is divided into two areas:

- research based on simulations for both transmission and distribution systems,
- analysis for real objects obtained from historical data.

2.1. Research Based on Simulations

The present literature is generally based on simulations. The analysis of power supply reliability in point of restoration issues was divided into two parts—distributed and transmission systems.

Distributed grid:

- The article [2] presents the networked microgrids aided approach to service restoration in a power distribution network. This paper proposes to use a mixed-integer linear approach. The main contribution of the article is to leverage networked microgrids to simplify service restoration. The proposed model was verified using the modified IEEE 123 node distribution test system.
- The article [3] deals with service restoration for a distribution network. The element under consideration is the uncertainty of restoration time. In the article, a two-stage adaptive algorithm for service restoration was proposed. This algorithm uses the Wasserstein distance metric. It is applied to calculate two restoration times with different probabilities. Then the higher probability is used as the restoration time.
- The paper [4] describes a multi-stage restoration method. It is applied to an medium voltage (MV) distribution system with distributed generation. The proposed service restoration approach concerns intentionally connection islanding of distributed generators (DGs) with network reconfiguration to maximize restoration of switched-off loads. It is realized by matching islanding schemes. Then the restoration of network connectivity and DGs is realized. Finally, the network reconfiguration as well as load shedding optimization are proposed. This research is based on a Pacific Gas & Electric (PG&E) sixty-nine bus system.
- The article [5] proposes a heuristic method for distribution network restoration. The proposed algorithm was implemented and tested on the IEEE 33-bus standard network.
- The article [6] concerns optimal network restoration after faults in a distribution network with distributed generation. The selected method is a meta-heuristic Artificial Bee Colony algorithm. The restoration algorithm and the load flow analysis were simulated using MATPOWER in MATLAB software. That research aimed to minimize out-of-service loads and power losses and improve the voltage profile. The article presents two examples of two single-fault and multi-fault cases. For each example, five different scenarios were studied. The results showed the significant power loss reduction and improvement in minimum voltage.
- Other papers that concern simulations to increase reliability are based on restoration issues for distributed grids, e.g., static island power supply restoration strategy [7], power restoration method using a genetic algorithm [8], the state-of-the-art fault localization and service restoration [9], robust power supply restoration for self-healing active distribution networks [10], intelligent power supply restoration [11], power restoration strategy [12], and a fast power service restoration method [13].

Transmission grid:

- The article [14] concerns power system restoration planning. The strategy presented in the paper uses an optimal energizing time needed to sectionalize islands. The method contains the identification of transmission lines that are not adequate to connect to the islands. The article methods consist of a combination of optimization methods: heuristic and discrete. The heuristic one is used to indicate an initial solution which is close to the optimal solution. Then it is input to the discreet method, which is the discrete Artificial Bee Colony approach.
- The paper [15] is based on a resilience analysis of transmission line restoration. It indicated that transmission line capacitance is based on resilience factors. The proposed ideas were verified in two IEEE tests.
- The article [16] presents a parallel automated resilience-based approach to restoration. The appliance aims to minimize the influence of the emergency power outages in a power system. The article proposes that during the power restoration process, a black start element is allocated to a little region on an as-needed demand. Then a mixed-integer nonlinear programming approach is indicated. The bi-level programming was used in the proposed solution to such a large-scale optimization model. The application was realized using both 6 and 118 bus IEEE test systems.
- The article [17] presents a possibility to solve the problem of expansion planning. The article contains the proposition of using multistage stochastic programming to solve this issue. The indicated mixed-integer linear programming proposes the placement of the construction and reinforcement of new transmission lines to assure the high reliability and quick restoration. The presented results are based on the IEEE 30-bus system with assuring to minimalize cost.
- The article [18] proposes the post-disaster restoration planning model that enables finding an optimal repair and activation schedule for damaged system components. In this model, an aim is to maximize load accommodation capability, as well as to minimalize the make-span of the restoration process. The obtained results increased maintenance efficiency. The IEEE 118 and 30 bus test systems were tested in the study. Moreover, the advantages of using the sequence-dependent repairing period are discussed.
- Other papers that concern simulation results in transmission systems and reliability are: using interline dynamic voltage restoration [19], a method for the optimization of a power system restoration path [20], a transmission line restoration using an emergency restoration system structure [21], an indication of the maintenance schedule of transmission lines [22], a definition of a restoration strategy in a transmission system during windstorm [23].

2.2. Research Based on Historical Data and Real Objects

The previous subsection includes literature resources from the last three to five years. However, all of them are based on simulations and different models (e.g., IEEE models). The authors indicate that there is a lack of present research of restoration issues based on real data, even if it is only input to further algorithms. This article is devoted to transmission lines restoration time; thus, this part is narrowed down only to transmission grids. Some interesting papers that concerns real object analysis using historical data can be found in the "SCOPUS" data base for key words "restoration" & "transmission lines" from the last three to five years:

- The article [24] presents a black start case study. However, the article contains simulations which are based on real data from Benghazi North Power Plant. The data were used to validate a black start plan for steady-state and transient operating conditions. The article indicates that the optimum size selection of the black start is defined by the capacity of the biggest motor, transmission line capacitive charging reactance, transformers size, and vector group.
- The article [25] presents a fault location system. The system is based on synchro phasors measurements. It is used for 345 kV and 161 kV transmission networks at Taiwan Power Company. Additionally, the article presents an evaluation based on historical cases.

- The paper [26] presents an issue that was connected with noticeable transmission lines failures in India under natural disasters. Data used in the article consisted of historical measurements when real disasters happened. The article discusses emergency restoration system applications. This system uses structure and foundation information, weather-related failure information, weather conditions, structural loading, and damage sizes.
- The paper [27] is related to the economic impact of climatic events in the USA. It additionally discusses why emergency restoration plans are needed. The second part of the article presents a case study from Oman. It presents emergency restoration procedures to downed transmission lines. Key aspects of emergency restoration procedures are discussed. The article indicates that with the development of materials and techniques, emergency restoration procedures must be periodically reviewed using actual technologies.
- Other papers that concern using real data in a transmission system and reliability are: an analysis of the empirical probability distribution of transmission line restoration time over 14 years [28], a case study of black starts of transmission lines in Australia [29], the development of a sequential restoration strategy and its empirical verification in a Korean power system [30].

2.3. Motivation and Contribution of the Paper

The number of articles, indicated in previous subsections, that concern simulation in recent years is huge. However, there is a lack of recent analyses realized for real cases although such real case analyses were common in the previous century. However, they are still necessary because the development of the materials and technics has totally changed in different areas. Thus, this article is a case study for the Russian power supply company "Oreolenergo" that concerns the analysis of restoration time in power lines based on analyzing historical data and a survey. In the analyzed regional power supply company "Oreolenergo", there are no monitoring systems for outages in the considered electrical networks. There is practically no automation equipment, and power lines are made radial. The structuring of the power supply restoration time given in the study makes it possible to consider in more detail all the constituents of the power supply restoration time and to establish the factors affecting it.

It is difficult to assess the real time of power supply restoration, since there is no real data on the time of the power supply outage beginning. The countdown of the power supply restoration time in most cases starts from the moment when the information on a failure is received by the dispatcher of a power supply company. As a result, restoration time is underestimated. Consequently, the damage from the undersupply of electricity is also inaccurate. Therefore, the constituent "time for obtaining information" was introduced into the structure of the power supply restoration time. The value of this constituent was determined by the Delphi method. This method was also used for the determination of the time for recognizing information. The choice of this method is due to the impossibility of evaluating the data on power supply restoration time constituents by other means. The questionnaires used for Delphi were designed specifically for this purpose and were sent to specialists who work in the power supply company "Oreolenergo" and who have at least five years of work experience. The choice of experts was justified by the fact that the dispatchers of the power supply companies receive information from consumers about power supply outages, register the moment of this information receipt, send a repair brigade to search for places of damage and to eliminate the identified damage, and register the corresponding time intervals for repairs and switching on. In most cases of power supply outages, especially in the 0.4 kV electrical network, dispatchers receive information about outages from consumers, since there are no monitoring systems for power supply outages. Nevertheless, they have cases of receiving information about the time of obtaining information, for example, during planned power supply interruptions made by the personnel of power supply companies. In these cases, they can register the time of disconnection and the moment of receipt of the information on the power supply outage from the consumer. Thus, there is the opportunity to analyze the time interval for obtaining information on power supply outages and the experts were competent in estimating time intervals for obtaining and recognizing information. A total of 20 experts responded to the questionnaires. In turn, the repair time was determined by using the method of mathematical statistics, while the time for connection harmonization was determined by using the analysis of literature sources. To summarize the method's end elements indicated for the restoration time indication, Figure 1 was prepared. Additionally, in the article, the technical means of electrical network automation to reduce the power supply restoration time were also found.

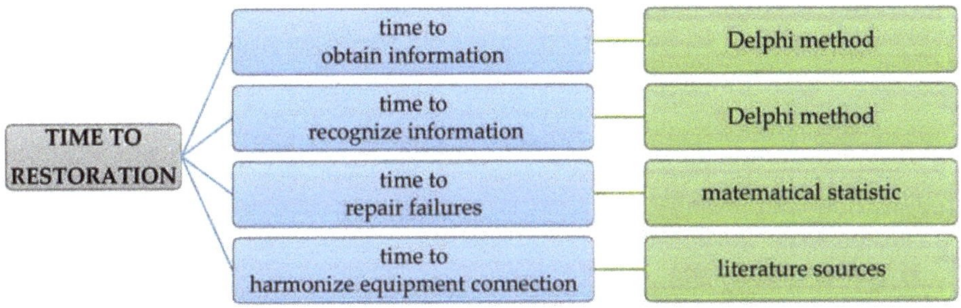

Figure 1. The proposed methodology to obtain restoration time for a selected power supply company.

3. Power Supply Reliability

One of the main criteria for power supply efficiency is power supply reliability, which implies the continuous supply of electricity to consumers in accordance with an electricity consumption schedule [31]. In turn, a main indicator of the PS reliability is power supply restoration time [32]. It can include the following constituents: time for obtaining information, time for information recognition, time to repair failures, and time for connection harmonization [33]. That is, the PS restoration time can be determined by Equation (1):

$$t_{\text{restor.}} = t_{\text{obt.infor.}} + t_{\text{rec.infor.}} + t_{\text{repair}} + t_{\text{harmonize}} \quad (1)$$

where

- $t_{\text{obt.infor.}}$—time to obtain information;
- $t_{\text{rec.infor.}}$—time to recognize information;
- t_{repair}—time to repair failures;
- $t_{\text{harmonize}}$—time to harmonize equipment connection.

Each component of this equation can be further analyzed and contains several more time intervals, each of which ultimately has an impact on the overall power supply restoration time.

A time for obtaining information is denoted as an interval from the beginning of a failure until obtaining information on it by a dispatching service of a PS company [33]. This time includes the following intervals:

$$t_{\text{obt.infor.}} = t_{\text{infor1}} + t_{\text{infor2}} + t_{\text{infor3}} \quad (2)$$

where

- t_{infor1}—time for obtaining information on failures by means of primary information links. This link can be electrical equipment receiving power energy from an electrical network and disconnecting in case of a power failure, a sensing device of an automation system, or network status monitoring (for example, a voltage sensor);
- t_{infor2}—time for obtaining information on failures by means of secondary information links. It can be the compared element of an automation system as well as a monitored network status. The specified time interval can be significantly reduced in the case of the use of automation, since a person (consumer) noticing disconnected equipment has to make sure that this disconnection occurred due to failures;

- t_{infor3}—time for obtaining information on failures by means of third information links. This link can be a dispatcher that receives a network failure signal or an element of a network status monitoring or another automation system making a decision based on received information (for example, a data processing unit, a microprocessor, etc.). This time interval largely depends on the data transmission channel. Thus, a person (consumer) can report a failure by phone, e-mail, or in person to the dispatcher, etc. t_{infor3} will be different in each of these cases.

The information recognition time may be described using this equation: [33]

$$t_{rec.infor.} = t_{read.infor} + t_{dec} + t_{search} + t_{report} \tag{3}$$

where

- $t_{read.infor}$—time required for information message recognition, that concerns failures in an electrical network. This time also depends on the data transmission channel through which the message arrived, the method of data transfer, and the speed of data recognition (who decrypts the message: a person or automatic equipment);
- t_{dec}—time spent on a decision by a dispatching office. It includes a time to decode information on failure, and it lasts until a place and a failure type are determined by a brigade;
- t_{search}—time required for a brigade to search the failure (depends on transport type, the remoteness of the failure place, the terrain type, the failure type, and brigade equipment for the search);
- t_{report}—time required to send information on a location and a failure type by a brigade (depends on the type of data transfer).

The repair time is an interval starting from the preparation of equipment to eliminate a failure up to the harmonization of the repair equipment [33]. This time can be represented as the following equation:

$$t_{repair} = t_{repair.prepar} + t_{repair.reach} + t_{repair.switch} + t_{repair.permit} + t_{repair.work} + t_{repair.complet} \tag{4}$$

where

- $t_{repair.prepar}$—time required for a repair brigade to depart including the preparation of work permit, equipment, devices, and loading on transport;
- $t_{repair.reach}$—time required for a repair brigade to reach a failure location. It depends on the distance to the failure place, the transport type, the landscape, road condition, the season, and the time of day;
- $t_{repair.switch}$—time required to switch necessary equipment;
- $t_{repair.permit}$—time required to obtain a permit for the work of a repair brigade. It depends on the work complexity as it impacts the preparation time of the workplace, that is, the implementation of technical measures to perform safe work;
- $t_{repair.work}$—time required to carry out direct repair work. It depends on brigade staff (quantitative and qualitative ones) and equipment with the appropriate tools and devices, along with the complexity of work;
- $t_{repair.complet}$—time required for the completion of work, the cleaning of a workplace, the exit of a repair brigade from a workplace, documenting the completion of work.

The time of the connection harmonization $t_{harmonize}$ can be described using this equation [33]:

$$t_{harmonize} = t_{inf.transfer} + t_{pre.connect} + t_{connect} + t_{ensure} \tag{5}$$

where

- $t_{inf.transfer}$—time required for information transfer time to a dispatching office the need to connect repaired equipment;

- $t_{pre.connect}$—time required to prepare the equipment connection and to document this;
- $t_{connect}$—time of equipment connection. It depends on the network diagram, the type of devices used for switching on, the distance from the personnel carrying out the switching up to the switching devices;
- t_{ensure}—time required to ensure that the equipment was successfully connected.

The literature positions indicate that data on the above time intervals are incomplete or often missed. However, the analysis of these time intervals reveals the potential to realize a reduction of the power supply restoration time that results in the power outages to consumers and the associated failures. Since the diagnostic methods and technical means for obtaining information about failures may be different [34], there are many factors that make it difficult to accurately determine the power supply restoration time and each of its constituents.

4. Results

4.1. Obtaining Information Time

The most correct method of determinization of obtaining information time of failures in electrical networks ($t_{obt.infor}$) is the Delphi method. This method was successfully applied in different researches, e.g., [35,36].

The questionnaire was prepared specifically for this research. It was given to twenty experts working in PS companies (dispatchers). The experts had at least five years employment experience.

This research proposes obtaining information time in 12 intervals. The specialists had to give a score from one to ten for each time interval. The most probable time interval got ten points from experts while the least probable got zero points. In the case that the expert indicates the same time interval probability, they could estimate the time intervals by points. The harmonization degree of the participant of the questionnaire was calculated. For this, a concordance coefficient (Equation (6)) proposed by Kendall was used:

$$W = \frac{12 \times S}{m^2 \times (n^3 - n)},$$
$$W = \frac{12 \times 35.9 \times 10^3}{20^2 \times (12^3 - 12)} = 0.62 \quad (6)$$

where

- S—the sum of squared differences between the sum of the estimates given by all experts to the i-th time interval ($\sum_{i=1}^{m} N_{ij}$) and the arithmetic mean of all the estimates $\overline{N}$;
- m—the number of experts surveyed; n is the number of time intervals in the questionnaire;
- N_{ij}—the score given by the j-th expert to the i-th time interval.

$$W = \frac{12 \times 35.9 \times 10^3}{20^2 \times (12^3 - 12)} = 0.62 \quad (7)$$

The arithmetic mean of all estimates was determined in accordance with the well-known Equation (8):

$$\overline{N} = \frac{\sum_{i=1}^{n} \sum_{j=1}^{m} N_{ij}}{n}$$
$$N = \frac{172 + 180 + 164 + 135 + 117 + 97 + 71 + 61 + 57 + 41 + 31 + 19}{12} = 95.4 \quad (8)$$

The sum of the squares of differences was determined according to Equation (9):

$$S = \sum_{i=1}^{n} \left(\sum_{j=1}^{m} N_{ij} - \overline{N}\right)^2$$
$$S = (5.8 + 7.1 + 4.7 + 1.5 + 0.5 + 0.00256 + 0.6 + 1.2 + 1.5 + 3 + 4.2 + 5.8) \times 10^3 = 35.9 \times 10^3 \quad (9)$$

Since the time intervals were indicated in the questionnaires, fixed points were chosen for calculating the expectation at each interval. These points corresponded to the middle of the intervals. The mathematical expectation was determined by the following equation:

$$M(t) = \frac{\sum_{i=1}^{n}(t_{ci} \cdot \sum_{j=1}^{m} N_{ij})}{\sum_{i=1}^{n}\sum_{j=1}^{m} N_{ij}}$$
$$M(t) = \frac{21.5 + 67.5 + 102.5 + 111.3 + 131.6 + 133.3 + 115.3 + 114.3 + 121.1 + 97.3 + 81.0 + 54.0}{1145} = 1.01 \quad (10)$$

where

- $M(t)$—the mathematical expectation of the time for obtaining information;
- t_{ci}—the time value of the middle of the i-th interval.

The calculation results are indicated in Table 1.

For clarity, the distribution of expert estimates given to the corresponding time interval is presented in Figure 2.

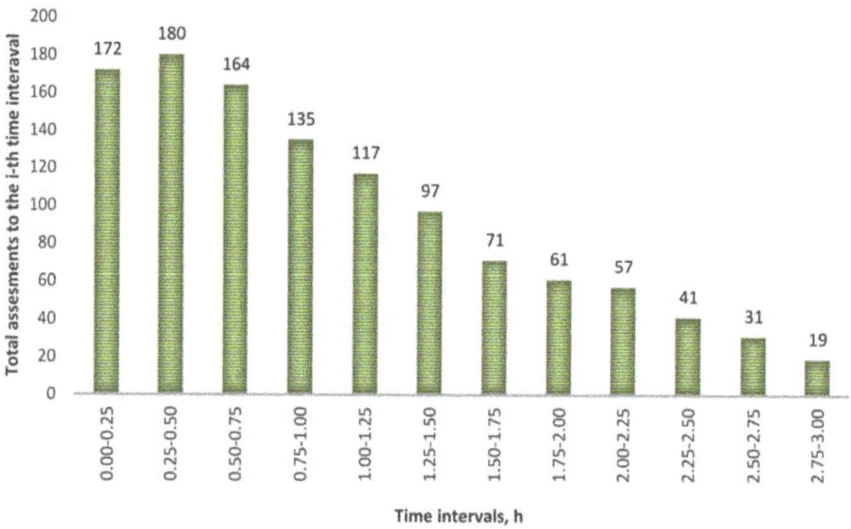

Figure 2. The histogram that represents the assessment of experts concerning a distribution of the time for obtaining information on failures.

The mathematical expectation of obtaining information time on failures was 1.01 h with the concordance coefficient of 0.627. In the questionnaires, it was considered that there were no monitoring systems of electric network, i.e., a PS company dispatcher obtained information on failures from the consumers.

This is quite a long time, which can and should be reduced by various means. A proposition to reduce the time may be, e.g., an automatic detection of failures facts and places in electrical networks and unmanned aerial vehicles allowing to monitor the power line state and detect failure places.

Table 1. The results of the expert survey to determine the time for obtaining information on failures.

Experts	Expert Estimates Given to the i-th Time Interval											
	Time Intervals, Hours											
	0.00–0.25	0.25–0.50	0.50–0.75	0.75–1.00	1.00–1.25	1.25–1.50	1.50–1.75	1.75–2.00	2.00–2.25	2.25–2.50	2.50–2.75	2.75–3.00
	1	2	3	4	5	6	7	8	9	10	11	12
1	9	8	8	7	8	4	3	3	2	3	5	2
2	8	10	9	6	5	5	2	3	3	2	1	3
3	10	9	7	8	6	4	3	4	3	1	2	1
4	7	9	8	7	6	5	4	2	3	3	2	1
5	9	10	7	8	4	3	5	3	2	1	2	2
6	8	8	9	6	7	5	3	4	2	2	1	1
7	10	10	7	5	6	4	3	3	3	2	2	0
8	8	10	9	6	5	5	2	3	2	2	1	0
9	10	9	7	8	6	4	4	3	2	3	2	1
10	7	8	9	6	5	4	2	3	3	2	2	0
11	8	9	9	7	5	6	3	2	1	3	0	0
12	9	9	8	6	7	5	5	3	4	2	1	1
13	8	7	6	7	5	4	2	3	4	2	1	1
14	9	8	10	7	6	5	4	2	3	1	1	0
15	9	10	7	8	8	6	4	3	3	2	1	1
16	8	8	10	6	7	6	5	5	4	2	2	2
17	7	10	9	8	6	6	5	4	3	3	1	0
18	10	9	7	7	5	6	4	3	4	2	2	1
19	8	9	10	6	5	5	4	2	3	2	1	1
20	10	10	8	6	5	5	4	3	3	1	1	1
$\sum_{i=1}^{m} N_{ij}$	172	180	164	135	117	97	71	61	57	41	31	19
t_{ci}	0.125	0.375	0.625	0.825	1.125	1.375	1.625	1.875	2.125	2.375	2.625	2.875
$t_{ci} \cdot \sum_{i=1}^{m} N_{ij}$	21.5	67.5	102.5	111.3	131.6	133.3	115.3	114.3	121.1	97.3	81.0	54.0
$\sum_{i=1}^{m} N_{ij} - \overline{N}$	76.6	84.6	68.6	39.6	21.6	1.6	−24.4	−34.4	−38.4	−54.4	−64.4	−76.4
$(\sum_{i=1}^{m} N_{ij} - \overline{N})^2$	5.8×10^3	7.1×10^3	4.7×10^3	1.5×10^3	0.5×10^3	2.6×10^3	0.6×10^3	1.2×10^3	1.5×10^3	3.0×10^3	4.2×10^3	5.8×10^3
	$W = 0.627$ $M(t) = 1.01$											

4.2. Recognizing Information Time

The present data indicated in the literature that concerns time of recognizing information on failures $t_{rec.infor}$ is not fully explored. However, it is worth noting that this time interval may take 75% of the time of the PS restoration. The Delphi method was also used to determine this time, and there were also 20 experts.

The results of the calculations are summarized in Table 2, the distribution of expert assessments given to this time interval is shown in Figure 3.

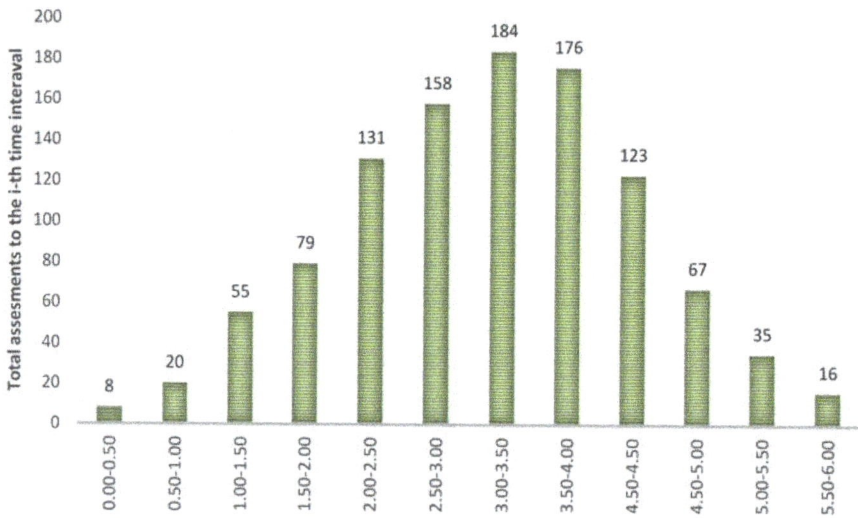

Figure 3. The histogram that represents the assessment of experts concerning the distribution of the recognizing information time.

The mathematical expectation of a time of recognizing information on failures was indicated as 2.94 h (a concordance coefficient is 0.79).

Table 2. The results of the survey of experts to determine the time for recognizing information on failures.

Experts	Expert Estimates Given to the i-th Time Interval											
	Time Intervals, Hours											
	0.00–0.50	0.50–1.00	1.00–1.50	1.50–2.00	2.00–2.50	2.50–3.00	3.00–3.50	3.50–4.00	4.00–4.50	4.50–5.00	5.00–5.50	5.50–6.00
	1	2	3	4	5	6	7	8	9	10	11	12
1	0	0	1	3	6	8	10	9	5	2	1	0
2	1	1	3	2	5	7	9	8	4	3	2	1
3	0	1	2	4	7	7	9	9	6	2	3	2
4	1	1	2	3	6	8	8	8	7	4	1	1
5	0	0	1	2	6	7	9	10	4	2	2	0
6	1	2	3	4	5	7	8	9	7	5	3	0
7	0	1	4	6	6	8	9	8	6	3	2	0
8	0	0	2	4	7	9	9	10	8	5	1	1
9	1	1	5	5	6	9	10	8	7	4	2	2
10	0	2	3	3	5	8	10	8	6	3	1	0
11	0	1	4	4	8	9	10	9	7	4	1	1
12	0	0	2	3	7	9	8	8	6	3	1	0
13	1	1	4	6	8	8	10	9	8	3	2	1
14	1	1	3	5	9	9	9	8	6	2	2	2
15	0	0	2	3	7	7	8	10	7	4	0	0
16	1	3	3	5	6	6	10	9	5	5	3	0
17	0	2	3	5	5	7	9	8	4	4	2	1
18	1	1	2	4	7	9	10	10	7	3	2	1
19	0	1	4	4	8	9	10	9	7	4	1	1
20	0	1	2	4	7	7	9	9	6	2	3	2
$\sum_{i=1}^{m} N_{ij}$	8	20	55	79	131	158	184	176	123	67	35	16
t_{ci}	0.25	0.75	1.25	1.75	2.25	2.75	3.25	3.75	4.25	4.75	5.25	5.75
$t_{ci} \cdot \sum_{i=1}^{m} N_{ij}$	2	15	68.75	138.25	294.75	434.5	598	660	522.75	183.75	81	92
$\sum_{i=1}^{m} N_{ij} - \overline{N}$	−79.8	−67.8	−32.8	−8.8	43.2	70.2	96.2	88.2	35.2	−20.8	−52.8	−71.8
$(\sum_{i=1}^{m} N_{ij} - \overline{N})^2$	6.4×10^3	4.6×10^3	1×10^3	0.0774×10^3	1.9×10^3	4.9×10^3	9.2×10^3	7.8×10^3	1.2×10^3	0.4×10^3	2.8×10^3	5.1×10^3
	W = 0.79 $M(t) = 2.94$											

4.3. Repair Time

This work uses the statistical data of time to repair failures t_{repair} obtained by using the year statistics of a power supply company ("Oreolenergo") [37]. The used data that concern failures and repair time are summarized in Table 3 and shown in Figure 4.

Table 3. Number of failures for specific repair time.

Repair time, h.	0.00–0.50	0.50–1.00	1.00–1.50	1.50–2.00	2.00–2.50	2.50–3.00	3.00–3.50	3.50–4.00
Number of failures, pcs.	75	50	29	38	12	3	2	3

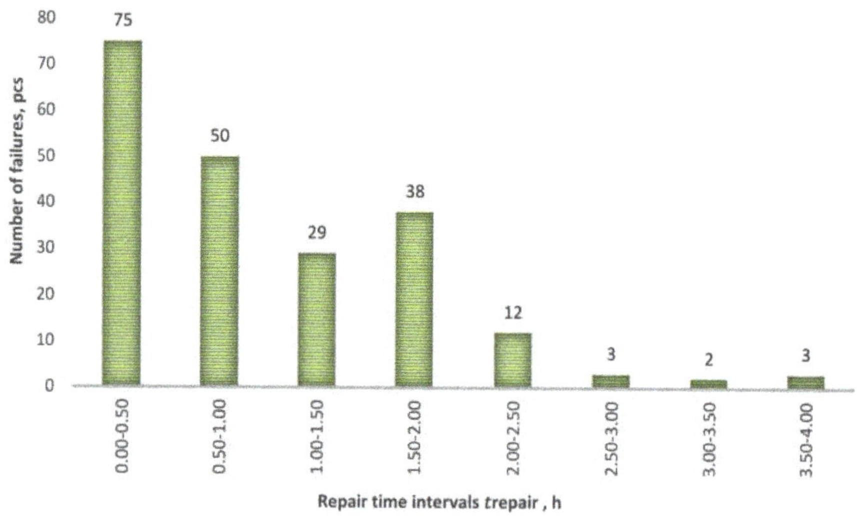

Figure 4. Number of failures for specific repair time.

The values of the mathematical expectation and dispersion of the repair time were determined from this data [5,10,18]. To do this, the sampling mean X_B was found out by Equation (11):

$$X_B = \frac{1}{n} \times \sum_{1}^{N_i} (X_i \times N_i)$$
$$X_B = \frac{0.25 \times 75 + 0.75 \times 50 + 1.25 \times 29 + 1.75 \times 38 + 2.25 \times 12 + 2.75 \times 3 + 3.25 \times 2 + 3.75 \times 3}{212} = 1 \quad (11)$$

where

- n—number of failures, $n = 212$;
- X_i—i-th repair time for which the calculation is made;
- N_i—frequency of the i-th time value.

Next, the conditional values U_i were determined by Equation (12):

$$U_i = \frac{X_i - C}{h}$$
$$U_1 = \frac{0.25 - 0.25}{0.5} = 0 \quad (12)$$

where

- C—constant (the repair time with the highest frequency of occurrence), C = 0.25 is for the first time interval;
- h—scale (the time step h = 0.5 h).

Similarly, other indicators were calculated, and the results are summarized in Table 4.

Table 4. Conditional values of the repair time.

No interval	1	2	3	4	5	6	7	8
U_i	0	1	2	3	4	5	6	7

The conditional sample value U_B was determined by Equation (13):

$$U_B = \frac{1}{n} \times \sum_{1}^{N_i} (U_i \times N_i)$$
$$U_B = \frac{0 \times 75 + 1 \times 50 + 2 \times 29 + 3 \times 38 + 4 \times 12 + 5 \times 3 + 6 \times 2 + 7 \times 3}{212} = 1.5 \quad (13)$$

The sample value X_B through the conditional sample value U_B was found by Equation (14):

$$X_B = U_B \times h + C$$
$$X_B = 1.5 \times 0.5 + 0.25 = 1 \quad (14)$$

The value of sample dispersion D_B was determined by Equation (15):

$$D_B = \frac{1}{n} \times \sum_{1}^{N_i} [(X_i - X_B)^2 \times N_i]$$
$$D_B = \frac{1}{212} \times [(0.25-1)^2 \times 75 + (0.75-1)^2 \times 50 + (1.25-1)^2 \times 29 + (1.75-1)^2 \times 38 + (2.25-1)^2 \times 12 + (2.75-1)^2 \times 3 + (3.25-1)^2 \times 2 + (3.75-1)^2 \times 3] = 0.61 \quad (15)$$

Root-mean-square deviation:

$$\delta_B = \sqrt{D_B}$$
$$\delta_B = \sqrt{0.61} = 0.781 \quad (16)$$

The corrected root-mean-square deviation S was found out to obtain a more accurate value of the deviation:

$$S = \sqrt{\frac{n}{n-1}} \times \delta_B$$
$$S = \sqrt{\frac{212}{212-1}} \times 0.781 = 0.782 \quad (17)$$

The probability of determining the repair time interval was taken to be $\gamma = 0.95$. Therefore, the value for determining the interval is $t = 1.96$.

The accuracy of the assessment:

$$2F = \gamma = 0.95$$
$$F = 0.475 \quad (18)$$

The estimation deviation:

$$\frac{t \times S}{\sqrt{n}} = \frac{1.96 \times 0.782}{\sqrt{212}} = 0.105 \quad (19)$$

The boundaries of the confidence interval:

$$X_B - \frac{t \times S}{\sqrt{n}} = 1 - \frac{1.96 \times 0.782}{\sqrt{212}} = 0.895 \text{ is the lower interval}$$
$$X_B + \frac{t \times S}{\sqrt{n}} = 1 + \frac{1.96 \times 0.782}{\sqrt{212}} = 1.105 \text{ is the upper interval} \quad (20)$$

Thus, the time to repair failures was in the interval 1 ± 0.105 h with 95% probability.

4.4. Connection Harmonization Time

The time for the harmonization of the equipment connection $t_{\text{harmonize}}$ depends on the applied communication tools, the time for preparing the equipment to be connected and its documentation, the time for equipment connection, and the time needed to ensure that the connection was successful.

The dispatcher must check the possibility of switching on a power line [38]:

- on records in the operational log and applications;
- by the telephone book "About the delivery and acceptance of lines";
- by the absence of posters on the drives of disconnectors;
- by interviewing operating personnel of substations and power plants about the absence of working people on the power line equipment which should be switched on.

After that, the command to turn on the equipment is given. In total, the time of connection harmonization can take up to 20 min.

4.5. Analysis Results

Considering the values of time intervals, the PS restoration time was calculated using Equation (1):

$$t_{\text{restor.}} = 1.01 + 2.94 + 1.00 + 0.33 = 5.28 \text{ h}$$

The indicated time of power supply restoration based on our analysis is equal to 5.28 h. This duration is significant and can cause considerable material damage to consumers supplied from the electric grid where blackout has occurred especially for those sensitive to process shutdown.

5. Discussion

Annually, the investigated regional power supply company "Oreolenergo", a branch of IDGC of Center, has an average of 344 power supply outages with a total of 98,495.835 kWh of unsupplied electric power [17]. The total number of power outages includes consumer outages, outages due to damage of overhead lines, cable lines, transformers, equipment of transformer substations, and distribution points. The average power supply restoration time is 5.86 h, which is close to the time obtained based on the performed studies (5.28 h). The deviation is 9.9%. It should be noted that these data on the power supply restoration time are related to failures in power transmission lines, which are the most unreliable element of the power supply systems.

A power supply restoration time of more than 5 h causes significant damage both to consumers and to power supply companies. Opportunities should be sought to shorten this time as much as possible. In most cases, it is difficult, since a noticeable number of power lines especially in rural areas have surpassed their resources and require replacement [39,40]. According to the statistical data of "Oreolenergo" [17], the causes of damage to power lines are shortcomings in maintenance (45%), the influence of natural and weather conditions (33%), the influence of unauthorized persons (15%), other reasons (e.g., birds, animals, etc.) (7%). At the same time, it is indicated that shortcomings in maintenance includes fallen trees and short circuits because of trees touching power line wires caused by untimely cleaning of power line routes, breaks of wires, fallen utility poles, and other causes associated with power line aging and late monitoring of their condition.

The PS restoration time may be reduced by different methods, especially the electrical network automation means. Almost all automation means can reduce the time to perform a particular operation and increase the accuracy of its execution. For example, the time to obtain information and recognize it can be significantly reduced by using a power supply reliability monitoring system or by using means for monitoring the technical condition of an electric network equipment, such as using unmanned aerial vehicles (UAVs) as in the articles [41–44] or the thesis [45]. Calculations realized in [45] indicated the implementation of the developed power supply reliability monitoring system in the Mtsensky electric grid of the Orel Region, Russian Federation. They showed that the time to obtain information was reduced from 1.01 h to 0.09 h, and the time to recognize information was reduced from 2.30 h to 0.25 h. In [46], UAV tests were described based on the Orelenergo branch of the IDGC of Center, PJSC. They showed that it was possible to achieve a reduction in the time of a PTL round check (the time for recognizing information) from 3.5 h/km to 5 … 15 min/km, that is, more than 30-fold.

It should be noted that if there are means of sectionalizing and redundancy of power lines such as automatic circuit reclosers (ACR), the time of power supply outages can also be significantly reduced [47,48]. However, in this case, the power supply restoration will be carried out by redundancy means, and the time for this restoration $t_{\text{restor.redund.}}$ is determined by the equation

$$t_{\text{restor.redund.}} = t_{\text{damaged section isol.}} + t_{\text{backup power act.}} \tag{21}$$

where $t_{\text{damaged section isol.}}$ is the time spent on the isolation of the damaged section from intact ones, h; $t_{\text{backup power act.}}$ is the time spent on backup power actuation, h.

All means of electrical network automation can be classified according to the time intervals that they shorten. This classification is shown in Table 5. Thus, the existing and promising methods and technical means of electrical network automation aim at reducing the specific constituents of the power supply restoration time.

Table 5. Methods and technical means of electrical network automation for reducing the power supply (PS) restoration time.

Time Interval	Methods and Technical Means of Electrical Network Automation
$t_{\text{obt.infor.}}$	• Monitoring of the technical condition of electric network equipment; • Monitoring of power supply reliability; • Telecontrol.
$t_{\text{rec.infor.}}$	• Monitoring of the technical condition of an electric network equipment; • Monitoring of power supply reliability; • Monitoring of electric network operation modes; • Telecontrol; • Means for determining the failure location.
t_{repair}	• Automation tools (processes digitalization of work authorization of brigades, registration of the beginning and end of work, etc., for example, the "Digital Electrician complex" [15]); • Repair work automation.
$t_{\text{harmonize}}$	• Means for remote communication of the brigade members with the dispatcher and with each other; • Monitoring the technical condition of electric network equipment; • Monitoring of electric network operation modes; • Telecontrol.
$t_{\text{damaged section isol.}}$	• Automatic sectionalizing of power lines; • Monitoring the technical condition of electric network equipment; • Monitoring of power supply reliability; • Telecontrol.
$t_{\text{backup power act.}}$	• Automatic redundancy of power lines and consumers; • Monitoring the technical condition of electric network equipment; • Monitoring of power supply reliability; • Telecontrol.

6. Conclusions

The power supply restoration time analysis of power transmission lines shows that it depends on several constituents. The constituents of time restoration analyzed in this study are:

- time for obtaining information,
- time for information recognition,
- time to repair failures,
- time for connection harmonization.

In this article, the methods of obtaining each of them were proposed and described. The case study calculations were realized for the Russian power supply company "Oreolenergo". The obtained restoration time was theoretically equal to 5.28 h. This value is equal to the statistical data obtained from the selected power supply company. The result deviation was less than 10%. Additionally, after obtaining the value of PS restoration time, it was proposed how it may be decreased. The proposition of the technical means of electrical network automation was indicated. These means were proposed and ordered in accordance to the time intervals they shorten.

Future research directions will be aimed at numerically estimating the impact of methods and technical means of electrical network automation on the constituents of the power supply restoration time. This will allow evaluating the effectiveness of their introduction by comparing the received values of the power supply restoration time constituents with initial ones and determining the reduction in damage from a lack of electricity supply.

Author Contributions: Conceptualization, A.V. (Alexander Vinogradov); methodology, A.V. (Alexander Vinogradov), and A.V. (Alina Vinogradova); validation, M.J. and T.S.; formal analysis, V.B., M.J., and E.J.; investigation, V.B., A.V. (Alina Vinogradova); resources, A.V. (Alexander Vinogradov); data curation, A.V. (Alexander Vinogradov); writing—original draft preparation, V.B.; writing—review and editing, M.J.; visualization, V.B., A.V. (Alina Vinogradova); supervision, T.S., Z.L., R.G.; project administration, A.V. (Alexander Vinogradov); funding acquisition, Z.L. All authors have read and agree to the published version of the manuscript.

Funding: This research received funding from the Chair of Electrical Engineering Fundamentals (K38W05D02), Wroclaw University of Technology, Wroclaw, Poland.

Conflicts of Interest: The authors declare no conflict of interest.

References

1. Liu, Y.; Fan, R.; Terzija, V. Power system restoration: A literature review from 2006 to 2016. *J. Mod. Power Syst. Clean Energy* **2016**, *4*, 332–341. [CrossRef]
2. Arif, A.; Wang, Z. Service restoration in resilient power distribution systems with networked microgrid. In Proceedings of the 2016 IEEE Power and Energy Society General Meeting (PESGM), Boston, MA, USA, 17–21 July 2016; pp. 1–5.
3. Li, J.; Song, X.; Wang, Y.; Zhang, X.; Tang, W. Service Restoration for Distribution Network Considering the Uncertainty of Restoration Time. In Proceedings of the 2016 3rd International Conference on Systems and Informatics (ICSAI), Shanghai, China, 19–21 November 2016; pp. 188–192.
4. Wang, F.; Chen, C.; Li, C.; Cao, Y.; Li, Y.; Zhou, B.; Dong, X. A Multi-Stage Restoration Method for Medium-Voltage Distribution System with DGs. *IEEE Trans. Smart Grid* **2017**, *8*, 2627–2636. [CrossRef]
5. Molaali, M.; Abedi, M. A New Heuristic Method for Distribution Network Restoration and Load Elimination Using Genetic Algorithm. In Proceedings of the 2018 Electrical Power Distribution Conference (EPDC), Tehran, Iran, 9–10 May 2018; pp. 46–51.
6. Gechanga, M.K.; Kaberere, K.K.; Wekesa, C. Optimal power service restoration using artificial bee colony algorithm. *Int. J. Sci. Technol. Res.* **2019**, *8*, 1950–1956.
7. Chen, Y.; Yang, P. Bi-level Programming Model for Distribution Generation in Active Distribution Network Considering Static Island Power Supply Restoration Strategy. In Proceedings of the 2019 22nd International Conference on Electrical Machines and Systems (ICEMS), Harbin, China, 11–14 August 2019; pp. 1–6.

8. Fan, Y.; Zi, X.; Jun, L.; Bingbing, L. An integrated power restoration method based on improved genetic algorithm for active distribution network. In Proceedings of the 2017 2nd International Conference on Power and Renewable Energy (ICPRE), Chengdu, China, 20–23 September 2017; pp. 524–528.
9. Mwifunyi, R.J.; Kissaka, M.M.; Mvungi, N.H. Distributed approach in fault localisation and service restoration: State-of-the-Art and future direction. *Cogent Eng.* **2019**, *6*. [CrossRef]
10. Yang, Q.; Jiang, L.; Ehsan, A.; Gao, Y.; Guo, S. Robust Power Supply Restoration for Self-Healing Active Distribution Networks Considering the Availability of Distributed Generation. *Energies* **2018**, *11*, 210. [CrossRef]
11. Jiang, L.; Yang, Q. Intelligent power supply restoration in power distribution networks with distubuted generation. In Proceedings of the 2016 China International Conference on Electricity Distribution (CICED), Xi'an, China, 10–13 August 2016; pp. 1–6.
12. Haishuang, G.; Liping, Q.; Demin, Y. Research on the Fast Power Restoration Strategy of Distribution Network Based on MAPSO. In Proceedings of the 2017 International Conference on Smart City and Systems Engineering (ICSCSE), Changsha, China, 11–12 November 2017; pp. 175–179.
13. Niu, G.; Zhou, L.; Qu, H.; Xiao, T.; Pei, W.; Qi, Z.; Kong, L. A fast power service restoration method for distribution network with distributed generation. In Proceedings of the 2017 IEEE Transportation Electrification Conference and Expo, Asia-Pacific (ITEC Asia-Pacific), Harbin, China, 7–10 August 2017; pp. 1–6.
14. Abu Talib, D.; Mokhlis, H.; Abu Talip, M.; Naidu, K.; Suyono, H. Power System Restoration Planning Strategy Based on Optimal Energizing Time of Sectionalizing Islands. *Energies* **2018**, *11*, 1316. [CrossRef]
15. Khalil, Y.; El-Azab, R.; Abu Adma, M.A.; Elmasry, S. Transmission Lines Restoration Using Resilience Analysis. In Proceedings of the 2018 Twentieth International Middle East Power Systems Conference (MEPCON), Cairo, Egypt, 18–20 December 2018; pp. 249–253.
16. Abbasi, S.; Barati, M.; Lim, G.J. A Parallel Sectionalized Restoration Scheme for Resilient Smart Grid Systems. *IEEE Trans. Smart Grid* **2019**, *10*, 1660–1670. [CrossRef]
17. Han, S.; Kim, H.-J.; Lee, D. A Long-Term Evaluation on Transmission Line Expansion Planning with Multistage Stochastic Programming. *Energies* **2020**, *13*, 1899. [CrossRef]
18. Yan, J.; Hu, B.; Xie, K.; Tai, H.-M.; Li, W. Post-disaster power system restoration planning considering sequence dependent repairing period. *Int. J. Electr. Power Energy Syst.* **2020**, *117*, 105612. [CrossRef]
19. Padmarasan, M.; Babu, R.S. Analysis of Interline Dynamic Voltage Restoration in Transmission Line. In *Intelligent Computing in Engineering. Advances in Intelligent Systems and Computing*; Springer: Singapore, 2020; pp. 587–595.
20. Kunlong, S.; Jianming, J.; Zongbao, G.; Gongyi, S.; Yue, L. An innovative method for optimization of power system restoration path. In Proceedings of the 2019 Chinese Control and Decision Conference (CCDC), Nanchang, China, 3–5 June 2019; pp. 4225–4230.
21. Prasad, D.; Khan, M.I.; Barua, P.; Agarwal, H. Transmission line restoration using ERS structure. In Proceedings of the 2017 14th IEEE India Council International Conference (INDICON), Roorkee, India, 15–17 December 2017; pp. 1–6.
22. Tian, N.; Dai, J.; Zhu, L.; Wang, Y.; Tan, Z.; Zhong, H.; Liu, B. Corrective short-term transmission maintenance scheduling considering post-contingency restoration. In Proceedings of the 2017 China International Electrical and Energy Conference (CIEEC), Beijing, China, 25–27 October 2017; pp. 699–703.
23. Gao, X.; Chen, Z. Optimal Restoration Strategy to Enhance the Resilience of Transmission System under Windstorms. In Proceedings of the 2020 IEEE Texas Power and Energy Conference (TPEC), College Station, TX, USA, 6–7 February 2020; pp. 1–6.
24. Asheibi, A.; Shuaib, S. A Case Study on Black Start Capability Assessment. In Proceedings of the 2019 International Conference on Electrical Engineering Research & Practice (ICEERP), Sydney, Australia, 24–28 November 2019; pp. 1–5.
25. Lin, T.-C.; Yang, J.-Z.; Yu, C.-S.; Liu, C.-W. Development of a Transmission Network Fault Location Platform Based on Cloud Computing and Synchrophasors. *IEEE Trans. Power Deliv.* **2020**, *35*, 84–94. [CrossRef]
26. Tamil Selvan, M.; Malar, K. Challenges faced for installation of emergency restoration system (ERS) in the major natural disaster hit high voltage transmission lines in India. *Disaster Adv.* **2019**, *12*, 1–8.
27. Vadivel, K.K. Emergency Restoration of High Voltage Transmission Lines. *Open Civ. Eng. J.* **2017**, *11*, 778–785. [CrossRef]

28. Kancherla, S.; Dobson, I. Heavy-Tailed Transmission Line Restoration Times Observed in Utility Data. *IEEE Trans. Power Syst.* **2018**, *33*, 1145–1147. [CrossRef]
29. Viawan, F.; Banktavakoli, R. Power System Restoration: Blackstart Studies. In Proceedings of the 2018 Conference on Power Engineering and Renewable Energy (ICPERE), Pittsburgh, PA, USA, 20–24 July 2018; pp. 1–6.
30. Goo, B.; Jung, S.; Hur, J. Development of a Sequential Restoration Strategy Based on the Enhanced Dijkstra Algorithm for Korean Power Systems. *Appl. Sci.* **2016**, *6*, 435. [CrossRef]
31. Drouineau, M.; Maïzi, N.; Mazauric, V. Impacts of intermittent sources on the quality of power supply: The key role of reliability indicators. *Appl. Energy* **2014**, *116*, 333–343. [CrossRef]
32. Baggini, A. *Handbook of Power Quality*; John Wiley & Sons: Hoboken, NJ, USA, 2008.
33. Vinogradov, A.; Vasiliev, A.; Bolshev, V.; Semenov, A.; Borodin, M. Time Factor for Determination of Power Supply System Efficiency of Rural Consumers. In *Handbook of Research on Renewable Energy and Electric Resources for Sustainable Rural Development*; IGI Global: Hershey, PA, USA, 2018; pp. 394–420, ISBN 9781522538677.
34. Vinogradov, A.; Vasiliev, A.; Bolshev, V.; Vinogradova, A.; Kudinova, T.; Sorokin, N.; Hruntovich, N. Methods of Reducing the Power Supply Outage Time of Rural Consumers. In *Renewable Energy and Power Supply Challenges for Rural Regions*; IGI Global: Hershey, PA, USA, 2019; pp. 370–392.
35. Turoff, M.; Linstone, H.A. The Delphi method-techniques and applications. *J. Mark. Res.* **2002**, *13*, 317.
36. Skulmoski, G.J.; Hartman, F.T.; Krahn, J. The Delphi method for graduate research. *J. Inf. Technol. Educ. Res.* **2007**, *6*, 1–21. [CrossRef]
37. IDGC of Center. *Technical Reports of the Branch of PJSC "IDGC of Center"—"Orelenergo", 2014–2016*; IDGC of Center: Moscow, Russia, 2016.
38. *Instruction on the Procedure for Operations during Disconnecting and Connecting 110 kV High-Voltage Power Lines Which are under the Management of a Power Company Dispatcher*; PJSC-IDGC of Center: Moscow, Russia, 2013.
39. Zeng, Y. Introduction to rural power grid problems and solutions. *Sci. Technol. Innov. Her.* **2013**, *7*, 138–140.
40. Harish, S.M.; Morgan, G.M.; Subrahmanian, E. When does unreliable grid supply become unacceptable policy? Costs of power supply and outages in rural India. *Energy Policy* **2014**, *68*, 158–169. [CrossRef]
41. Jones, D. Power line inspection—A UAV concept. In Proceedings of the IEE Forum on: Autonomous Systems, London, UK, 28 November 2005; Volume 2005, p. 6.
42. Li, Z.; Liu, Y.; Hayward, R.; Zhang, J.; Cai, J. Knowledge-based power line detection for UAV surveillance and inspection systems. In Proceedings of the 2008 23rd International Conference Image and Vision Computing New Zealand, Christchurch, New Zealand, 26–28 November 2008; pp. 1–6.
43. Zhou, G.; Yuan, J.; Yen, I.-L.; Bastani, F. Robust real-time UAV based power line detection and tracking. In Proceedings of the 2016 IEEE International Conference on Image Processing (ICIP), Phoenix, AZ, USA, 25–28 September 2016; pp. 744–748.
44. Chernyshov, V.A.; Semenov, A.E.; Bolshev, V.E.; Belikov, R.P.; Jasinski, M.; Garifullin, M.S. The method of extending drone piloting autonomy when monitoring the technical condition of 6–10 kV overhead power lines. In Proceedings of the E3S Web of Conferences 2019, Kazan, Russia, 18–20 September 2019; Volume 124, p. 02010. [CrossRef]
45. Bolshev, V. Development of Technical Means for Monitoring Power Supply Outages and Voltage Deviation at Inputs of Rural Consumers. Ph.D. Thesis, Federal Scientific Agroengineering Center VIM, Moscow, Russia, 2019.
46. IDGC of Center. *Protocol of the Test of an Unmanned Aerial Vehicle for Inspection of Overhead Lines on the Basis of Mtsensky "IDGC of Center"—"Orelehnergo"*; IDGC of Center: Moscow, Russia, 2014.
47. Abiri-Jahromi, A.; Fotuhi-Firuzabad, M.; Parvania, M.; Mosleh, M. Optimized Sectionalizing Switch Placement Strategy in Distribution Systems. *IEEE Trans. Power Deliv.* **2012**, *27*, 362–370. [CrossRef]
48. Vinogradov, A.; Vinogradova, A.; Bolshev, V.; Psarev, A.I. Sectionalizing and Redundancy of the 0.38 kV Ring Electrical Network. *Int. J. Energy Optim. Eng.* **2019**, *8*, 15–38. [CrossRef]

energies

Article

A Case Study on Distributed Energy Resources and Energy-Storage Systems in a Virtual Power Plant Concept: Technical Aspects

Tomasz Sikorski [1], Michal Jasiński [1,*], Edyta Ropuszyńska-Surma [2], Magdalena Węglarz [2,*], Dominika Kaczorowska [1], Paweł Kostyla [1], Zbigniew Leonowicz [1], Robert Lis [1], Jacek Rezmer [1], Wilhelm Rojewski [1], Marian Sobierajski [1], Jarosław Szymańda [1], Daniel Bejmert [3], Przemysław Janik [3] and Beata Solak [1]

1 Faculty of Electrical Engineering, Wroclaw University of Science and Technology, 50-370 Wrocław, Poland; tomasz.sikorski@pwr.edu.pl (T.S.); dominika.kaczorowska@pwr.edu.pl (D.K.); pawel.kostyla@pwr.edu.pl (P.K.); Zbigniew.Leonowicz@pwr.edu.pl (Z.L.); robert.lis@pwr.edu.pl (R.L.); jacek.rezmer@pwr.edu.pl (J.R.); wilhelm.rojewski@pwr.edu.pl (W.R.); marian.sobierajski@pwr.edu.pl (M.S.); jaroslaw.szymanda@pwr.edu.pl (J.S.); beata.solak@pwr.edu.pl (B.S.)

2 Faculty of Computer Science and Management, Wroclaw University of Science and Technology, 50-370 Wrocław, Poland; edyta.ropuszynska-surma@pwr.edu.pl

3 TAURON Ekoenergia Ltd., 58-500 Jelenia Góra, Poland; daniel.bejmert@tauron-ekoenergia.pl (D.B.); przemyslaw.janik@tauron-ekoenergia.pl (P.J.)

* Correspondence: michal.jasinski@pwr.edu.pl (M.J.); magdalena.weglarz@pwr.edu.pl (M.W.)

Received: 6 May 2020; Accepted: 8 June 2020; Published: 15 June 2020

Abstract: The article presents calculations and power flow of a real virtual power plant (VPP), containing a fragment of low and medium voltage distribution network. The VPP contains a hydropower plant (HPP), a photovoltaic system (PV) and energy storage system (ESS). The purpose of this article is to summarize the requirements for connection of generating units to the grid. Paper discusses the impact of the requirements on the maximum installed capacity of distributed energy resource (DER) systems and on the parameters of the energy storage unit. Firstly, a comprehensive review of VPP definitions, aims, as well as the characteristics of the investigated case study of the VPP project is presented. Then, requirements related to the regulation, protection and integration of DER and ESS with power systems are discussed. Finally, investigations related to influence of DER and ESS on power network condition are presented. One of the outcomes of the paper is the method of identifying the maximum power capacity of DER and ESS in accordance with technical network requirements. The applied method uses analytic calculations, as well as simulations using Matlab environment, combined with real measurement data. The obtained results allow the influence of the operating conditions of particular DER and ESS on power flow and voltage condition to be identified, the maximum power capacity of ESS intended for the planed VPP to be determined, as well as the influence of power control strategies implemented in a PV power plant on resources available for the planning and control of a VPP to be specified. Technical limitations of the DER and ESS are used as input conditions for the economic simulations presented in the accompanying paper, which is focused on investigations of economic efficiency.

Keywords: virtual power plant; distributed energy resources; energy storage systems; grid codes; power systems; smart grids; prosumer; business model; economic efficiency; sensitivity analysis

1. Introduction

A Virtual Power Plant (VPP) is still an actual approach and there is not a standardized definition for the framework of a VPP in the literature. The origin of the terminology of "Virtual Power Plant" may be traced back to 1997, when S. Awerbuch, in the book "The Virtual Utility—Accounting, Technology and Competitive Aspects of the Emerging Industry" defined Virtual Utility as flexible cooperation of independent, market-driven actors that assures an efficient energy service expected by the consumers without the need for appreciating assets [1]. A VPP manages distributed energy resources (DER) named also distributed generation (DG) units [2]. For example, wind, solar and hydroelectric power generation units are interconnected. Managing them together enables them to be more effective [3–5].

A Virtual Power Plant, as an autonomous, intelligent unit equipped with effective and safe power flow control systems, consists of generators, loads and energy storage that is connected to the distribution network [6]. These devices are equipped with controllers, which usually power electronic converters that allow for power flow control [6]. Coordinating the work of the entire VPP is a difficult and demanding task. The system's architecture must not only enable power flow control but also ensure VPP protection—not only related to power system security but also cybersecurity. In Reference [7], VPP architecture based on a common information model (CIM) and IEC standard 61850 is shown.

There have recently been many attempts to integrate intelligent solutions in power systems. An interesting discussion related to microgrids and the VPP is presented in Reference [8]. Microgrids allow increasing the efficiency of the use of distributed energy sources and energy storage systems. It also allows for regulating the load. Microgrids can be connected to the power system or operate as a standalone system. VPP management is based on computer software that enables the integration of distributed sources. In systems connected to the distribution network, the value of the power generated by the generation sources, the operation of the energy storage and the response of the demand side are optimized. Several propositions proclaiming the idea of transforming microgrids to a virtual power plant have recently been discussed, among others in References [2,6]. Additionally, it is also worth mentioning a new topic called virtual microgrids, which can be recognized as software solutions and algorithms supporting the planning, design and operation of microgrids. As an example of the virtual microgrids, it is worth noting a prosumer cluster connection into virtual microgrids to obtain cost reduction [9] or energy peer-to-peer trading in virtual microgrids [10].

Due to the random nature of the generated power, a large number of independent renewable energy sources can lead to system stability problems and therefore the connecting of distant generation sources, loads and energy storage units into a VPP has many benefits [11]. Work [12] shows the possibility of using charging points for electric vehicles, as well as wind generation, in the VPP concept. It also presents power flow optimization while taking into consideration price, wind generation and electric vehicles. Paper [13] concerns VPP control power consumption for heating. The operation algorithm is based on the application of thermal mass to the building to defer power consumption from electric space heating.

There are many different aspects to VPP power flow control. The main goals correspond to economic aspects related to electrical energy trading. The VPP control algorithms predict energy storage charging at low energy prices, as well as discharging energy storage and the sale of energy at peak demand at high prices. Paper [14] presents a stochastic bi-level optimization model to maximize day-ahead profit and to minimize predicted real-time production and the consumption of imbalance charges. In Reference [15], the bidding strategy of a VPP is determined using mathematical models. Ref. [16] presents decentralized coordination of VPP units, considering both active and reactive power using the novel Lagrangian relaxation-based mechanism. The method takes into account the effect of flexible demand and prevents the creation of new demand peaks and troughs. Another aspect concerns optimizing the use of locally generated energy and using the right strategy for storing energy in energy storage [17,18]. Power flow in a VPP, due to technical aspects, can also be optimized. In Reference [19], the binary backtracking search algorithm (BBSA) is used to optimize power flow in a VPP in order to achieve a reduction of generation cost and power losses, as well as, to increase reliability. To achieve the same goal, risk-constrained stochastic programming is used in Reference [20] and the Imperialist competitive algorithm is used in Reference [21]. A big problem in the modern generation can also be seen to be carbon dioxide emissions. In Reference [22], binary particle swarm optimization (BPSO) is used to solve the indicated issues.

However, technical issues cannot be overlooked when planning the different strategies for a VPP. For example, voltage levels at all points in the distribution network should be within the range allowed by the standards. The same applies to the values of currents in the transformer lines and windings. Cooperation between units included in the VPP, meeting these expectations, is presented in Reference [23]. Moreover, issues regarding the operation of the storage itself are also important. Studies on the impact of energy storage parameters on VPP strategy and performance are presented in References [24,25]. The crucial technical aim of the VPP is concentrated on the aggregation control of the number of distributed generation units, which are grid-connected close to consumers (end-users, households). The aggregation verification may be a centralized or not system supported by a logic control algorithm, as well as, a communication infrastructure [26]. The control strategies must concern reliability, uncertainty, stability, demand response, power quality, active and reactive power management, protection and balancing and reliability in various load circumstances [27–29].

Additionally, when technical aspects are indicated it is worth noticing the management entities—virtual power players issues. Virtual power players' aim is the generation and services remuneration or charging energy consumption. The diversity of players to expedite participation in the electricity markets is described in the literature [11,30,31]. Reference [32] proposes the methodology to DER management. The article includes resource scheduling, aggregation and remuneration. The aggregation process is realized by k-means algorithm. Clustering is realized for different approaches, that concerns tariffs definitions for the period of a week. Customer remuneration is realized in accordance to Portuguese time-of-use tariffs. The research corners twenty thousand consumers and five hundred distributed generation units. The paper [33] deal with the same issues. However, it is realized for 2592 operation scenarios. Those cases consider over 5 hundred DGs, over 20 thousand consumers and ten suppliers. The article [34] is another example of using clustering to prosumers aggregation. The article [35] presents the discussion of demand response in point economic pros and effectiveness. This article presents a sensitivity analysis of demand response prices for the virtual power player operation costs. Additionally, the analysis comparison of cost of distributed resources and demand-side response during facing supply unavailability. This calculation is performed in a real smart grid on buses with associated micro-production. This allows the creation of sub-groups of data according to their correlations. The clustering process is evaluated so that the number of data sub-groups that brings the most added value for the decision-making process is found, according to players' characteristics. In addition to the technical aspects, selected issues concerning the roles of VPP partners are discussed in the accompanying paper [36]. Physical and financial streams between them are highlighted in point of the decision model which is concentrated on profits maximization. The results show that the number of distributed energy resources and the available storage capacity of battery energy storage has an impact on the economic efficiency of the VPP.

This article aims to study the technical aspects of integration of the above-mentioned units with an electrical power system (EPS) with regards to their prospective application in the VPP. The mentioned limitations are related to the regulation and protection procedures applied in the control of generation units and storage systems and also their possible influence on power system parameters. This article presents calculations and power flow of a real virtual power plant (VPP), containing a fragment of low and medium voltage distribution network. The model contains the hydropower plant (HPP), a photovoltaic system (PV), energy storage system (ESS). The problem is based on the identification of the limitations, which are dictated by the technical requirements of cooperation of power generation units and energy storage systems with the power system; the application of the simplified calculation, which is supported by precise simulation techniques so that the maximum power capacity of the planned energy storage system in the planed VPP can be indicated and the investigation of real measurements of a photovoltaic power plant in order to reveal the impact of power control strategy on the potential of the resources integrated with the VPP. The presented review of VPP definitions and aims, as well as a summary of VPP projects, are the motivation for this paper. Additionally, to obtain the economic issues and impact on electricity marked firstly the technical requirements must be assured. The contribution of this paper covers the current gap of knowledge related to the VPP project which exhibits in the real case limitations of utilization the DER and ESS. For example, this paper provides the real case example of calculation and simulation focused on the determination of maximum power capacity of the ESS planned In terms of VPP efficiency and sensitivity, it is important to identify the maximum level of ESS power capacity that can be connected in the planned node to be installed in the selected network node. Results of presented investigations formulate margin condition for the VPP resources. Another example of the contribution of the paper to the current knowledge gap is the attempt to determine the influence of the power regulation strategies applied in PV plants on the real power range available for the VPP control strategies. Using the real measurement investigations it was shown that reactive power consumption implemented in the PV inverters reduces the energy volume potentially available for VPP form the PV installations. Mentioned examples constitute the limitation for the VPP project and can be adapted to other VPP investments.

The aim of this paper is to identify possible limitations in the development of the VPP which might be related to the regulation, protection and integration of power generation units with power systems. The mentioned problem can be seen to be a crucial issue, especially on the preliminary stage of the VPP concept and when different approaches for the economic strategies for VPP are created. After the introduction section, which highlights the main motivations of this paper, Section 2 presents a literature review of the technical aspects of VPP. This includes recent investigations regarding network integration, as well as a review of a selected real case of VPP realization. The aim of Section 2 is to identify current problems and solutions related to VPP and to provide the range of functionality of current VPP projects. Section 3 highlights the problems regarding the investigations presented in this paper. It has to be emphasized that technical aspects of DER and ESS, associated with the VPP, can be considered on several levels. Thus, Section 3 consists of the identification of codes and technical standards that define the requirements and permissible limits of electrical power system parameters and power quality, protection schemes and active and reactive power control issues applied for distributed energy resources. An additional element described in Section 3 is the energy storage control and limitations coming from the charging and discharging characteristics of energy storage systems. The revealed aspects can be treated as boundary conditions for the identification of their impact on VPP planning and operation strategies. The main investigations are presented in Section 4, which starts with a description of the topology of a medium voltage network in the area of the investigated VPP. The investigations are based on a real VPP project and present results considering the investigation of the cooperation of a 1 MW hydropower plant with 0.5 MW battery energy storage connected to the same node of medium voltage distribution network and impact of their operating condition on power flow and voltage level in observed network belonging to the VPP, (b) identification of maximum power capacity of battery energy storage which can be connected to considered node of the VPP as well as identification of general grid capacity of the investigated fragment of the distribution network to connect possible DERs/ESSs, (c) the identification of the impact of power control strategy applied in a PV power plant on resources available for the VPP. The maximum power capacity of the ESS is understood as the rated power of the ESS determined by requirements for power quality parameters of the grid and requirements for the integration of the generation units with power systems. In presented investigations, the storage capacity of the considered ESS is fixed and is used as a margin condition for the simulations. The storage capacity determines the ESS operability usually reflected by available time for charging and discharging. The selection of storage capacity of the ESS is usually based on specific aim functions considering selected intentions of using the ESS like economic aspects or islanding mode. The maximum power capacity of the ESS is restricted by the standards and regulations addressed to cooperation with the grids. The presented analysis is based on simulations conducted in Matlab combined with real measurements. The initial condition for the calculation was based on the real measurements of load and generated power that represent a day of summer load peak demand. As a result of the investigation, the maximum power capacity of the considered ESS is identified with regards to the requirements for the permissible level of rapid voltage changes. Additionally, the impact of the power control strategy applied in a PV power plant on resources available for the VPP was also identified. In Section 5 a discussion about the influence of requirements for grid connection applicable to power generation units and its impact on limitation of the maximum power capacity of distributed energy resources and energy storage systems considered for planning operation of the VPP is provided in the broadest context. Section 6 presents the conclusions.

The identified limitations for the VPP, resulting from the technical aspects presented in this paper, are used as the input conditions for the economic investigations presented in the accompanying paper [36]. The paper presents the results of economic efficiency, including sensitivity analysis on price factors and DER production volumes, as well as the capacity of ESS.

2. Literature Review in the Context of the Technical Aspects of the VPP

2.1. Selected Investigations of the VPP: Load Demand Reduction, Voltage Control, Islanding, Microgrids, Power Quality

In recent years, in the literature related to the technical aspects of a VPP, several examples are worth noting.

- Reference [37] introduces a regulation mechanism for a VPP. The calculation method of a virtual power plant's frequency performance is presented. The model proposed in the article enables the VPP's regulation performance score to be analyzed and the VPP's regulation control strategies to be simulated. The obtained results show that the strategy can reduce a VPP's variability caused by DGs.
- Reference [38] presents the possibility of using μ-CHP generation units in households, which would lead to a situation where consuming households will also be able to produce electricity. This enables the local management of the grid. The application of information and communication technologies (ICT) enables the clustering of μ-CHPs in VPP. The research was conducted by the companies ECN and Gasunie. The results indicated that a cluster of 10 households with a μ-CHP may reduce the substation peak load by 30–50% without infringing user comfort.
- Reference [39] indicates the need for efficient voltage control. The article presents the possibility of using small run-of-the-river hydropower plants, which are connected to a VPP to control the network voltage. The control is realized with the management of the reactive and active power of communicated hydro power plants belonging to a virtual power plant. The paper contains research on the efficiency of various voltage control measures. The small hydro power plant's active and reactive power enables the voltage in the electrical network to be controlled with PV during times of high feed-in.
- Reference [18] presents research on the optimal configuration model of an energy storage system (ESS) in a VPP with large-scale distributed wind power. The optimal objective function of the energy storage system is established with consideration of economy, load shifting and safety standards. The particle swarm optimization algorithm is used to solve the model. The model feasibility is verified on the IEEE 33 node system. The obtained results indicate that larger ESS configurations lead to a positive impact on load shifting.
- Reference [40] contains simulations of a VPP. The analyzed VPP consists of a 200 MW wind power plant (WPP), a 100 MW photovoltaic power plant, PV and +/− 250 MW pumped storage. The indicated DG units are integrated into an islanded grid with a thermal power plant. The base islanded grid load is 1300 MW. The article describes the control strategies of the pumped storage power plant. The analyzed strategies are focused on the improvement of the power quality (PQ) level in case of significant solar power variations.

- Reference [41] concerns the frequency control issues connected with an increasing amount of wind generation, which can be seen as an important part of a VPP. The authors indicate that the inertia of the system has an important function because it determines the influence on frequency variation during the changes in generation or power demand level. Thus, the doubly-fed induction generator wind turbines, which reduce the effective inertia of the system, may be used to control the frequency.
- Reference [42] contains a technical-economic impact analysis of the massive integration of small generators and demands into a VPP. The results can be observed in system functioning and on the outcome of demands and generators within the VPP. The paper contains the analysis and comparison of several VPP strategies. Additionally, the article proposes optimization algorithms based on the modification of the big band and big crunch (BB–BC) optimization method. The algorithms under research aim to determine the optimal location and optimal load control strategy of renewable energy sources and also the optimal operation schedule of energy storage elements in order to minimalize the energy purchased from a substation. The important outcome of the research is that a high reduction of energy purchased from substation energy is possible using the control of the load demand in a VPP.
- Reference [43] contains the description of a flexible energy optimizer for microgrids and VPPs. In the article, the unit commitment and economic dispatch problem is solved by an enhanced mixed integer linear programming (MILP) method. Additionally, different post-optimization modules were developed, which enabled the potential network constraint violations to be mitigated and power quality to be improved. One of the strategies is toward to enhance voltage and loading quality, reduce power losses and support selected ancillary service in worst voltage quality nodes and nodes with high power consumption.
- Reference [44] presents the architecture for a VPP and the interaction of customer meters using a virtual power plant controller. The paper contains the description of human machine interface (HMI) development to access reactive power metering at the location of customers. Additionally, the article presents a recording tool for reading a VPP controller. The VPP controller is able to control the reactive power flows to the customers and it, therefore, carries out a proactive operation to reduce voltage instability.
- Reference [45] contains the impact analysis of harmonic distortion on the energization of energy distribution transformers integrated into a VPP. One of the parts presents the analysis of an analytical procedure that predicts the inrush current and the parameters of a single-phase power transformer under a distorted voltage condition. The aim of the indicated analysis is that variations in inrush current resources correspond to the voltage harmonic distortion level expressed by the total harmonic distortion in the voltage. Although, it is important to notice that in large-scale home appliance use of electronic power equipment so the level of distortions may increase in the future.

2.2. *Review of Selected Real Case VPP Realizations*

Despite the many investigations concerning VPP concepts, strategies and control, it is worth noticing actual VPP projects, which show the scale of developments and results:

- The Fenix project was developed to improve the contribution of distributed energy resources to the electric network by aggregating them into Large Scale Virtual Power Plants. The project is based on the cooperation of DG owners, energy companies, research institutes and universities—EDF, Areva, Siemens, ECRO SRL, Imperial College London, the University of Manchester and the University of Amsterdam, among others. The project consists of three phases [46]. The first phase involves the preparation of two scenarios—northern and southern—which are described in this subsection. The analysis presented in paper [46] included the DER contribution to the network, as well as its strengths and weaknesses. The second phase covered the design of the communication and control system between the DERs in a VPP. The last phase concerned the validation of previously prepared analysis and systems through the realization of building 2 installations—in the UK and Spain [47]. Close to Bilbao City in Spain, there is a FENIX Southern Demonstration VPP connected to the Iberdrola system. The indicated installation connect a lot of different DER, such as combined heat and power (CHP), wind turbines, hydropower, photovoltaics, a CHP-biomass system, with a combined total capacity of 0.168 GW. The distribution area, that the project is realized has the peak demand equal to 0.320 GW. It works as a casual VPP. It means, that information from every connected distributed energy source is analyzed in the main control system. The information is exchanged in an intelligent interface called "FENIX Boxes." This system connects every DG. The indicated intelligent electronic devices assures adequate steering and implementation for the communication protocols. The protocols use wireless communication approaches for the distributed energy sources and control systems. It is based on the Virtual Private Network. The Virtual Private Network unites the heat and power plants, wind and solar sources and heat pumps, which assures creation of the interconnected, flexible system that has centralized control.

- The project Smart Power Hamburg aims to design a VPP. It is towered to aggregate the variable load and CHP units in Hamburg in Germany. The created Virtual Utility applies the present urban infrastructure. The infrastructure consists of CHP units, heat storage systems and a building with demand-side management. It presents a new direction to the generation of electricity, as well as, covering heat demand in the urban area. The connection of cogeneration and heat storage systems enables the unit to be operated according to electrical demand instead of heating demand, thus increasing the unit's flexibility. The project shows the possibility of integrating the bunkers storage possibility, swimming pools or heating networks, which are steered by the indicated virtual power plant [48].
- The Danish EDISON virtual power plant case was started in 2009. It is related to analyzing the Bornholm Island that provides supportive services to real energy market players for example, generation companies to obtain the effective application of distributed energy resources [49].
- A battery storage virtual power plant placed in the Australia. The South Australia conducted the biggest case study project. The indicated virtual power plant consists of a lot of the small-scale batteries and photovoltaics. The total capacity of indicated units is five MW (seven MWh). The indicated systems are connected to the central monitoring and management element in the VPP. The important element of the project is to simplify local network constraints, stabilize prices of electric energy and support renewable energy sources. The South Australia has great potential for exploitation of renewable energy—more than forty percent of the generated power proceeds from the wind farms or rooftop photovoltaic systems [50].

3. Problem Statement

The cited examples show significant progress in the development of VPPs. However, this article aims to highlight technical aspects at the preliminary stage of the concept and operation planning of a VPP. The size of the ESS, the capacity of available power generation, as well as control restrictions on the use of the DER and ESS have to be identified. The cited examples are also of interest due to permanent amendments of standards and network codes. This is why the problem has its beginnings in the technical requirements of the cooperation of power generation units and energy storage systems with a power system. The aim of this section is firstly to generally review the technical requirements dedicated to controlling the DER and ESS in terms of their cooperation with the power system and then to select and discuss in detail several topics which can appear on the primary level of VPP planning. Among the wide list of technical aspects, this paper presents three of them. They were selected due to the association of the VPP project with the distribution network and small generation units. Thus, the technical aspects presented in this section serve as the base for further investigation of the following problems:

- cooperation of the DER and ESS and their impact on the power flow and voltage level in the observed network belonging to the VPP,
- identification of maximum power capacity of ESS which can be connected to the considered node of the VPP and identification of grid capacity to connect possible DERs and ESSs
- investigation of the impact of the power control strategy applied in DERs on resources available for the VPP.

3.1. Review of Control of DER, Protection Systems, Requirements for the Connection to the Grid and Parallel Operation with EPS

Distributed energy resources, as well as energy storage systems, should fulfill the requirements in relation to relevant power system parameters including:

- the frequency and voltage levels, which are usually associated with the requirements for reactive power regulation, as well as frequency control issues,
- the short-circuit current contribution,
- the fault-ride-through capability, which exhibits itself in the requirements for protection devices and settings.

The issues of the impact of distributed energy resources on power system parameters were described by Conseil International des Grands Réseaux Électriques (CIGRE) in Reference [51] and Verband der Elektrotechnik Elektronik Informationstechnik (VDE) in Reference [52]. The European Commission established the requirements for connecting power sources to the grid in Reference [53]. The Agency for the Cooperation of Energy Regulators (ACER), in Reference [54], made it mandatory for transmission system operator (TSO) and distribution system operator (DSO) to include the mentioned requirements in network codes (NC). The network codes set out the necessary minimum standards and requirements that need to be followed when connecting the DER. A prominent example of the network code is the European Network of Transmission System Operators for Electricity (ENTSO-E) network code [55]. In the case of distributed generation with a connection point below 110 kV, the ranges of permissible limits of power system parameters are defined separately for four types of power generating modules, depending on the maximum power capacity. For the Central Europe networks, the different types of maximum capacity threshold from which a power generating module is a categorized are:

- "type A": 0.8 kW–1 MW,
- "type B": 1 MW–50 MW,
- "type C": 50 MW–75 MW,
- "type D": higher than 75 MW.

For the particular types of power generating modules at the point of connection to the grid, the following are defined:

- a permissible range of frequency,
- the active power frequency response capability that regards a limited frequency sensitive mode (overfrequency), as well as the maximum power capability reduction with a falling frequency concerning the limited frequency sensitive mode (underfrequency—selected parameters consider active power range, frequency response insensitivity, frequency response deadband, droop),
- pre-fault, post-fault circumstances for the fault-ride-through capability (voltage and time parameters),
- voltage and reactive power profiles in relation to the level of the actual value of the ratio of reactive and active power for consumption and generation scenarios (the characteristic coordinates of the (U–Q)/Pmax profile consists of the maximum range of Q/Pmax, which demonstrates the maximum range of cosϕ or tgϕ regulation, as well as the maximum range of the steady-state voltage level).

The mentioned types of power generating modules are subject to predefined compliance simulations. Additionally, the network operators define the scheme and settings that are necessary to protect the network, taking into account the profiles of the generation units. The network code [55] mentions the following aspects of protection schemes: " [55]

- external and internal short circuit,
- asymmetric load (negative phase sequence),
- stator and rotor overload,
- over/under excitation,
- over/under voltage at the connection point,
- over/under voltage at the alternator terminals,
- inter-area oscillations,
- inrush current,
- asynchronous operation (pole slip),
- protection against inadmissible shaft torsions (for example, subsynchronous resonance),
- power generating module line protection,
- unit transformer protection,
- backup schemes against protection and switchgear malfunction,
- overfluxing (U/f),
- inverse power,
- rate of change of frequency,
- neutral voltage displacement."

All the mentioned aspects have been listed in order to highlight the technical aspects which should be taken into consideration when VPP concepts are created. The presented paper considers a case study of a VPP concept which uses resources that are localized in the distribution network. Some general studies about standards related to small generators are presented in Reference [56]. In order to highlight the details of the control strategy applied for small generators in the following subsections, selected issues are described in detail, including active power reduction strategies and voltage profile requirements. The described elements are used in the further investigations, which concentrate on the identification of the maximum capacity of the network for the energy storage connection and assessment of the losses of active power capacity in a photovoltaic power plant regarding reactive power control strategy.

3.2. *Active Power Control Strategy Applied in DERs*

When highlighting the active power control strategy implemented in small generators, it is worth pointing out existing standards, such as VDE-AR-N-4105 [52]. When referring to this standard, the active power frequency response characteristic *P(f)* and standard characteristic of the Q/P_{max} (known also as *cosφ(P)* or *tgϕ(P)*) for the generation unit directly connected to a low voltage (LV) network can be revealed. These characteristics are presented in Figures 1 and 2, respectively. The power frequency response characteristic assumes that for power system frequency between 50.2 Hz and 51.5 Hz. When the exceedance of frequency is detected, it is recommended to reduce active power generation from P_M with a gradient of 40% P_M per Hertz. At system frequencies higher than 51.5 Hz and lower than 47.5 Hz, the power generation unit will be disconnected immediately.

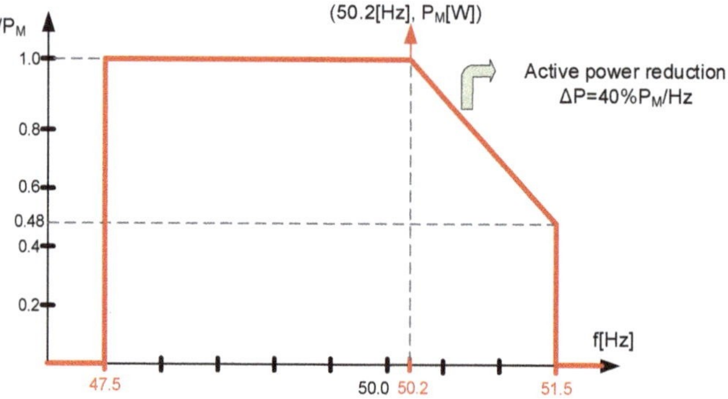

Figure 1. Standard active power frequency response characteristic *P(f)* applied for DER units integrated with an low voltage (LV) electrical network (based on Reference [52]).

When discussing relations between active power and reactive power, the characteristic of the $Q/P_{\max}$ (known also as $cos\varphi(P)$ or $tg\phi(P)$) for the generation unit can be introduced. This characteristic is presented in Figure 2. It can be concluded that up to 20% of the maximum power capacity $P_{\max}$, both generation and consumption of reactive power is recommended. In the range of (0.2–0.5) $P_{\max}$, the active power generation is accepted. When exceeding half of $P_{\max}$, both generations of the active power and reactive power consumption are recommended. It may be used as a model with the capacitive power factor $cos\varphi_{cap}$. The reactive power level consumption relays on a range of $P_{\max}$. The application of reactive power consumption aims to reduce the increase in voltage that is because of the noticeable active power generation. levels.

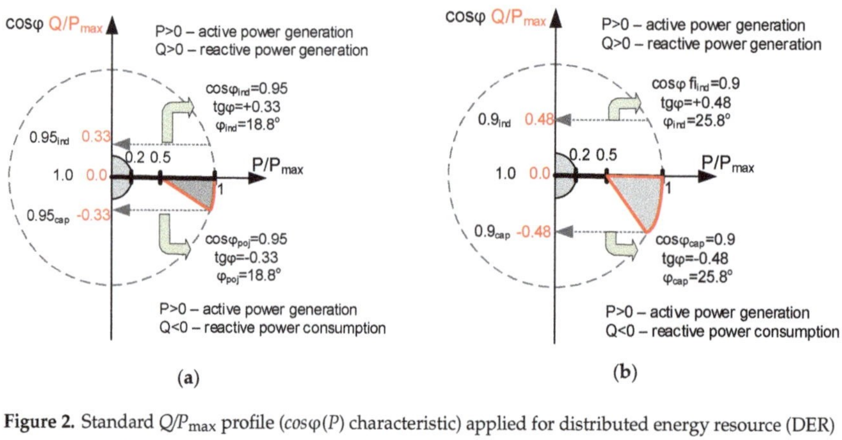

Figure 2. Standard $Q/P_{\max}$ profile ($cos\varphi(P)$ characteristic) applied for distributed energy resource (DER) units integrated with a LV electrical network: (**a**) with regard to units with a maximum power capacity of 3.68 kVA to 13.8 kVA; (**b**) with regard to units with the maximum power capacity over 13,800 VA (based on References [52,57]).

3.3. Power Quality Voltage Profile

A common area that can be related to the mentioned control and regulation strategies and protection schemes, is the power quality. The most crucial power quality parameters are frequency variations, voltage variation, voltage fluctuation, voltage unbalance (asymmetry), voltage harmonics, interhamonics, subharmonics, direct current injection (DC) and rapid voltage changes [58,59]. A general standard related to public electrical networks is EN 50160 [60]. In conjunction with quoted standards related to the DER, it is possible to define permissible limits of voltage changes in the connection points of the DER, as well as in other nodes of the network.

One of the critical requirements states that rapid voltage changes caused by switching operation with maximum power capacity cannot be more than 3% of the nominal voltage.

$$d_c = \frac{|\Delta U_c|}{U_a} 100 \leq 3\%, \quad (1)$$

where:

- ΔU_c—steady state voltage change
- U_a—nominal voltage
- d_c—relative steady state voltage change as a parameter of rapid voltage change

Additionally, a slow voltage changes (voltage level) in every node of distribution network consisting of DERs cannot exceed 10% of nominal voltage considering every DERs working simultaneously with maximum power capacity for 10-min aggregating time recommended by standard EN50160 [60]. Recently, in the last update of the mentioned standard [60], measurements with an aggregation time of 1-min instead of 10-min were suggested. Some investigations of the influence to aggregation interval on the assessment of photovoltaic power plant belonging to discussed VPP topology has been discussed in Reference [61].

3.4. Protection Schemes

In addition to the control and regulation characteristics, the aspects of protection schemes should also be considered. Table 1 consists of the over/under voltage and over/under frequency protection schemes at the connection point of DERs that are connected to an LV electrical network. The schemes are required by the DSO in selected European countries on the basis of References [52,56].

Table 1. Over/under voltage and over/under frequency protection schemes at the connection point of DERs connected to a LV electrical network in selected European countries.

Country	Under Voltage	Over Voltage	Under Frequency	Over Frequency
Germany	0.7–1.0 U_N; $t \leq 0.2$ s	1.0–1.15 U_N; $t \leq 0.2$ s	47 Hz; $t \leq 0.2$ s	52 Hz; $t \leq 0.2$ s
Italy	0.8 U_N; $t \leq 0.2$ s	1.2 U_N; $t \leq 0.1$ s	49–49.7 Hz; immediately	50.3–51 Hz; immediately
Spain	0.85 U_N; $t \leq 1.2$ s	1.1 U_N; $t \leq 0.5$ s	48 Hz; $t \leq 3$ s	51 Hz; $t \leq 0.2$ s
Belgium	0.5–0.85 U_N; $t \leq 1.5$ s	1.06 U_N; immediately	49.5 Hz; immediately	50.5 Hz; immediately
Poland	0.85 U_N; $t \leq 1.5$ s	1.15 U_N; $t \leq 0.2$ s	47 Hz; $t \leq 0.5$ s	51 Hz; $t \leq 0.5$s

3.5. Control of EES, Charging and Discharging Characteristics

Among energy storage devices, chemical batteries are increasingly used in professional power engineering [62]. The desirable features of batteries are their high energy density, high charge and discharge power and long life cycle. Other aspects are also relevant, such as the methods of determining the state of charge (SoC) and the possibilities of recycling [63]. For this reason, lithium-ion batteries are the most commonly used in battery energy storage (BES). The advantages of this type of battery include the fact that they have an energy density of 160 Wh/kg, a power density of up to 350 W/kg and a lifetime greater than 1000 charging and discharging cycles. The disadvantages of a lithium-ion battery include its high cost and the need for a heating and cooling system.

The main issue connected with controlling energy flow to and from an energy storage device is the correct determination of its operating characteristics [64]. Characteristics are defined by the manufacturer and they depend on the storage design. Furthermore, the system operator can control the storage operation using them. The rate of charging or discharging energy storage (also called C-rate) when using lithium-ion batteries can be determined as the value of the current at which the energy storage is discharged [65]. It is often expressed as the ratio of the battery capacity to the time of discharge. For example, a discharge rate of 1C means that storage will be completely discharged in 1 h. On the other hand, a discharge rate of 0.5C means that the same storage will be completely discharged in 2 h. [66] The storage charging or discharging rate is also determined by the remaining energy. The relative value of the remaining energy in relation to the rated capacity is called the state of charge (SoC) [67]. The characteristics of the dependence of the storage charging and discharging power on the SoC level should be provided by the manufacturer. The operator can change the shape of the characteristics within certain limits, for example, by preferring quick or slow charging or discharging in a specific SoC range [68]. In some cases, such shaping of characteristics can optimize storage efficiency and increase its lifetime and safety.

An important aspect of modeling energy storage operation is its lifetime and the decrease in capacity when using the battery. Reference [69] describes the impact of the ambient temperature and depth of discharge on the wear and tear and degradation cost of storage.

To ensure that each cell operates correctly within a certain voltage, temperature and current range during charging and discharging, the battery requires a built-in controller that communicates with the battery management system (BMS). The power value regulated by the BMS takes into account both the technical limitations of the technology and the safety conditions of the storage. The BMS is designed to maintain the efficient operation of the storage. The control is based on the current state of battery operation, that is, state of charge (SoC), temperature, counted discharge cycles and so forth.

The storage charging and discharging rate are especially affected by:

- Design—when designing batteries, manufacturers need to choose the size, weight, cost, lifetime and performance of the storage. Depending on needs, storage power and capacity can be lower due to cost and weight.
- State of Charge (SoC)—when the battery is almost fully charged, the charging speed decreases. The reason is that BMS prevents the cells from overheating. At 80–90% SoC, the charging speed usually drops significantly and slows down to almost zero at 100% SoC. Charging speed is most effective between 0% and 80% SoC.
- Temperature—lithium-ion battery cells work most effectively in the 20–30 degree Celsius range. When the battery temperature is too low or too high, the BMS reduces the current in order to protect the battery cells. If the storage is equipped with a heating and cooling system, the BMS controls the temperature of the storage cells by thermal management of the battery. The temperature of the battery depends on the ambient temperature and on the value of the charging or discharging power.

For the selected type of storage system, the dependence of charging and discharging power on the degree of SoC (modified Ragone plot) can be determined. The speed of charging (discharging) is determined by the power and is expressed in Watts or as a relative value in relation to the nominal power of the container. However, the SoC can also be defined as energy in Watt-hours or as a relative value in relation to the nominal capacity of the storage tank. Exemplary charging (discharging) of typical storage based on lithium-ion batteries is shown in Figure 3. An appropriate sign of the battery power was assumed for charging (positive-red) and discharging (negative-blue). The presented characteristics are typical for the lithium-ion batteries. Characteristics express the limitations of available power depending on the current SoC level. In discussed VPP, the storage unit has a power of 500 kW and is composed of lithium-ion batteries. It requires individual investigations to reveal the "real" charging/discharging characteristic, which may differ from that presented in Figure 3.

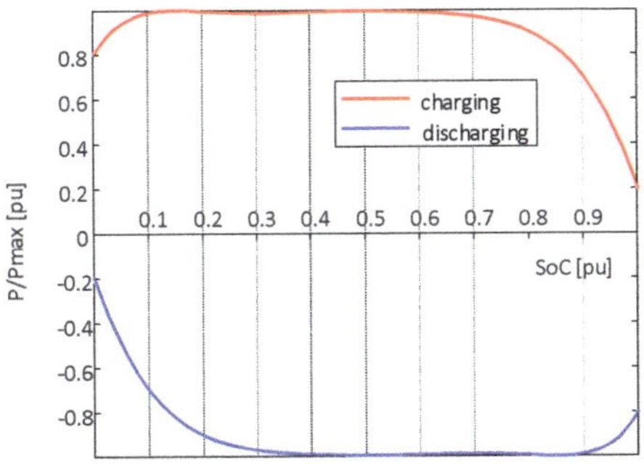

Figure 3. Dependence of charging and discharging power from the state of charge (SoC): Charging (positive-red) and discharging (negative-blue).

4. Investigation of the Impact of the DER and ESS on VPP Planning and Operating Conditions—A Case Study

Discussed technical aspects related to the impact of DERs on the operating condition of the network or power control strategy applied in DERs are primarily addressed to the integration of DER with the power systems. However, the revealed aspects can be treated as boundary conditions in terms of VPP planning and operation strategies. This section aims to present methods that can be used to solve three formulated in the problem statement topics. These issues are investigated based on a real VPP project. The obtained results are associated with the investigated localization of the VPP however can be treated as an example of a method for the assessment of formulated topics. These topics are as follows:

- Investigation of the cooperation of a 1 MW hydro power plant with 0.5 MW battery energy storage that is connected to the same node of medium voltage distribution network and also the impact of their operating conditions on the power flow and voltage level for analyzed network belonging to the VPP,
- identification of maximal power capacity of battery energy storage which may be connected to considered node of analyzed VPP as well as identification of general grid capacity of the investigated fragment of the distribution network to connect possible DERs/ESSs,
- identification of the impact of power control strategy applied in a PV power plant on other DER available for the VPP.

4.1. Description of the Case Study—Topology of the Planned VPP

The planned virtual power plant is based on the fragment of the distribution network in Poland. The topology of the VPP on the distribution network scheme is presented in Figure 4. The VPP area consists of two parts of distribution networks supplied from two HV/MV main stations—110/20 kV *R-J* and *R-Z*. The supplied stations are connected to the 110 kV electrical power system (EPS). The 20 kV network, fed from *R-J* station, is an overhead-cable network. The 20 kV network, fed from *R-Z* station, is mainly an urban cable network. Both networks work with earth fault current compensation.

Inside the mentioned distribution networks, there are several distributed energy sources and energy storage systems. Planned VPP consists of hydro power plant, photovoltaic power plant, biogas generation units and combined heat and power unit based on combined installation using boiler and steam turbine integrated with generator and heating system in the industry. An integral element of planed VPP is the prosumers mainly using photovoltaic systems. Detailed information about the DERs identified in the area of the VPP is presented in Table 2. The detailed information about ESSs localized in the VPP area is presented in Table 3. In the first part of the distribution network supplied from the station *R-J*, a crucial element of the planned VPP is the hydro power plant denoted as *HPP-L* with generating power about 0.94 MW and battery energy storage system *ESS-L* connected to the same node of MV network as hydro power plant with installed power 0.5 MW. In the second part of the distribution network supplied from the station *R-Z*, a photovoltaic power plant *PV-C* with generating power of 0.1 MW and an associated energy storage system can be noted.

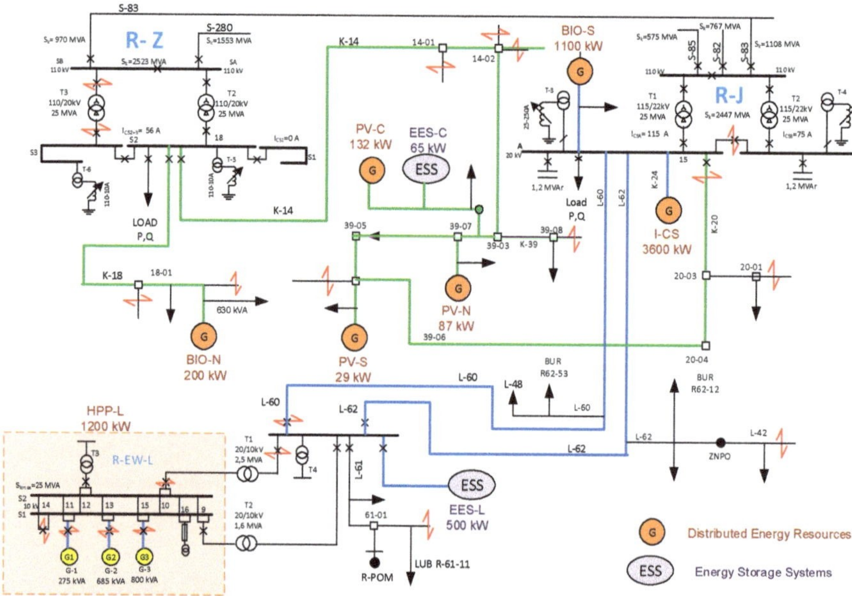

Figure 4. Topology of the planned virtual power plant (VPP).

Table 2. Renewable source of energy (RES) in the area of the planned VPP.

Name	Type	Installed Power [MW]	Generating Power [MW]	Connection Voltage Level
HPPL-L	hydro	1.250	0.94	MV
PV-C	photovoltaic	0.132	0.100	LV
I-CS	industrial CHP	3.960	3.600	MV
BIO-S	biogas	1.100	1.100	MV
BIO-N	biogas	200.000	150.000	MV
PV-S	photovoltaic	0.029	0.029	LV
PV-N	photovoltaic	0.087	0.087	LV
PV-MI	photovoltaic	microinstallation		LV

Table 3. Energy storage system (ESS) integrated with the VPP.

Name	Nearest RES	Installed Power	Connection Voltage Level
EES-L	hydro	500 kW/500 kWh	MV
EES-C	photovoltaic	65 kW/65 kWh	LV

4.2. *Identification of the Impact of the Active Power Changes Generated by the DER and ESS on the Load Reduction in the Distribution Line, as well as the Voltage Changes in the Nodes of the Grid Covered by the VPP*

In order to illustrate the technical aspects related to the integration of DER and ESS with electric power systems and also their impact on resources available to the VPP control, the issue of identifying the maximum power capacity of ESS is presented with regards to network requirements concerning voltage level and rapid voltage changes. The network requirements were described in Section 2. This paper aims to reveal the maximum capacity of the ESS that is planned to be attached to the same node as analyzed HPP. The investigated approach does not consider the valuable aspects of the interaction of ESSs with different RES at respective locations or analyze the advantages and disadvantages of using one ESS vs multiple small ESSs. However, these mentioned issues have been investigated in point of the economic aspects and are presented in associated paper [36]. The presented investigations are performed using Matlab modeling integrated with a database of real measurements of power flows. Referring to the VPP topology presented in Figure 4, the appropriate model was created and simulations were carried out in order to identify:

- the impact of switching on and off the active power generation of hydro power plant *HPP-L* on the power flows and on the voltage level in indicated distribution network considered in the VPP topology,
- the impact of a gradual increase of active power fed into the network jointly by hydro power plant HPP-L and energy storage system *ESS-L*, which is done to determine the maximum power capacity of *ESS-L* which is permissible from the point of view of rapid voltage changes parameter. The rapid voltage changes parameter is caused by the sudden shutdown of these energy sources.

Firstly, in order to associate the model with the real operating conditions of the distribution network associated with the VPP a steady state initial condition of power flows was prepared. Power flows were prepared based on the available set of annual real measurements of load demands and power generated by DER. Loads consist of the two main HV/MV transformers and all busbars of the power stations *R-J* and *R-Z*. One day of measurements, representing summer maximum peak demand, was selected. The day was selected based on the analysis of the database of the Polish power system load, which is shared by the Polish Transmission System Operator [70]. The presented investigations concern the day 14.07.2017 at 1 PM. A simplified scheme of the electrical connection of the VPP, with denoted instantaneous power flow measurements at the time of the identified summer peak demand value, is presented in Figure 5.

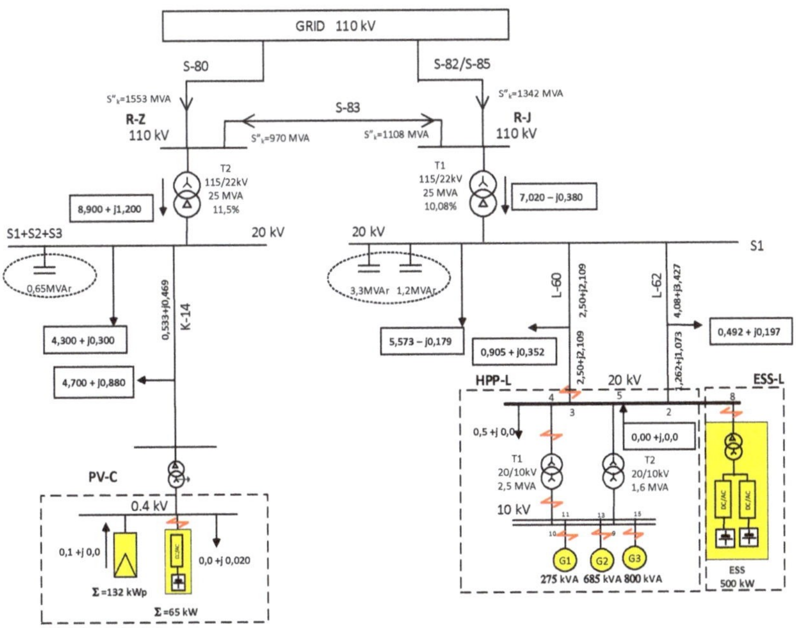

Figure 5. A simplified scheme of the electrical connection of the VPP, with denoted instantaneous power flow measurements at the time of the identified summer peak demand value.

Secondly, a Matlab Simulink model of the VPP topology, consisting of the described 20 kV networks, hydro power plant, battery energy storage and photovoltaic power plant was created and validated. The model is the dynamic model and captures several aspects related to dynamic representations of DERs and BES. In the case of HPP simulation, a standard electro-hydraulic speed controller model was used. The mechanical time constant and the time of waterfall was calculated based on the real parameters and equals 2 s and 3 s, respectively. The static excitation system was chosen according to the IEEE type ST1A excitation system mode in Matlab. The dynamics of the system are determined by the parameters of the voltage regulator and equalizer. Both elements are represented by an inertial element with the following parameters:

- regulator: $G_r(s) = \frac{K_a}{1+T_a s}$, with assumed parameters: $G_r(s) = \frac{210}{1+0.100s}$
- equalizer: $G_f(s) = \frac{K_f s}{1+T_f s}$, with assumed parameters: $G_f(s) = \frac{0.001s}{1+s}$.

In the case of the battery energy storage system, a functional modeling assumption was made that BES works like a controllable source of active and reactive power. The phenomena and processes occurring in the cells as well as in the control and commutation system of the inverter are not taken into consideration in the applied model. Applied limits are connected to restrictions on the discharge and charge current, the battery charge level and the power change speed. An ideal *P,Q* inverter regulation system, controlled by voltage magnitude and phase, was applied. Simulation of the BES operation in the grid was realized in accordance to the selected scenario, for example, determining the time intervals for energy return and battery charging. In addition, it is necessary to determine the power of exchange with the grid (discharge and charge current) and to also control the battery charge level. For this purpose, the *P,Q* inverter model was supplemented with a battery charge control system. The condition for the simulations assumed the use of the BES that was planned in the VPP project, with nominal parameters of 0.5 MW of maximal power $P_{\max}$ and a 0.5 MWh of maximum capacity. Additionally, a limitation for the speed of power change in the simulations was implemented to ± 10%$P_{\max}$/s. The numerical values given in the figures refer to BES with a 0.5 MW power and a 0.5 MWh useful capacity. The PV power plant is also modeled using the *P,Q* inverter model. However, in the performed simulations of short time intervals, fixed values of active and reactive power were used on the basis of real measurements collected for the investigated PV-C. In the applied model of PV-C, the issue of generation changes due to radiation and temperature changes are neglected. Initial condition for the simulation was supported by the real measurement data which represents selected summer peak demand.

Simulation time of 24 min (1440 s) was selected, which allowed all the assumed events in the simulation scenario, while at the same time maintaining the real dynamics of energy sources and storage systems during switching operations, to be performed. Time of the simulation results from the time interval of control and regulations systems. In the simulation model the issue of water turbine response, mechanical constant, limitation of the speed of power changes in the control of BES have been implemented. The particular time interval associated with the planned switching operations, carried out in the generation units and energy storage system, are defined in the scenario of the simulated events. The simulations use the algorithm for solving differential Equations known as ode24tb, which works with a variable integration step. The maximum integration step was 10^{-4} s, while the actual step was selected automatically. The accuracy of the timestamp is not worse than 10^{-4} s.

With regards to the scheme of the electrical power network in the area of the planned VPP (Figure 4), the presented simulations are focused on the active power changes and voltage changes of particular VPP elements, including:

- the hydropower plant *HPP-L*, which is connected in the distribution network that is supplied from the main station *R-J*,
- the battery energy storage, which is connected to the same node as the hydro power plant *ESS-L*,
- the main transformer *T1* in the main power station *R-J*,

- the distribution line *L-62* associated with main statin *R-J* and hydropower plant *HPP-L* with battery storage system *ESS-L*. *ESS-L* serves for supplying the energy consumers and for power transmission from the *HPP-L* and *ESS-L* to the main station *R-J*,
- the photovoltaic power plant *PV-C*.

The scenario of events consists of several switching operations carried out in the generation units and energy storage system:

- 5th s—switching on the hydro power plant *HPP*-L with presets: active power generation P_G = 940 kW and reactive power Q_G = 0.0 kVAr,
- 5th–900th s (15 min)—time interval for the HPP-L to reach the setpoints of the HPP-L control and regulation systems *HPP-L*,
- 900th s (15th min)—switching on the photovoltaic power plant *PV-C* with parameters: active power generation P_G = 100 kW and reactive Q_G = 0.0 kVAr,
- 1000th s (approx. 17th min)—switching on the full load battery energy storage *ESS-L* in the discharge mode with presets: active power generation P_G = 500 kW and reactive power Q_G = 0.0 kVAr,
- 1200th s (20th min)—switching off the hydropower plant *HPP*-L,
- 1300th s (approx. 22nd min)—switching off the battery energy storage *ESS-L*,
- 1400th s (approx. 23rd min)—switching off the photovoltaic power plant *PV*-C,
- 1440th second (24th min)—end of the simulation.

Active power changes in the distribution line *L-62*, associated with main statin *R*-J and hydropower plant *HPP*-L with battery storage system *ESS-L* during series of switching operations of *HPP-L*, *PV-C* and *ESS-L* are presented in Figure 6. It can be seen that the gradual increase of active power generated by the hydro power plant *HPP-L* from zero to the assumed setup of 940 kW generating power decreases the load demand in line L-62 which connects *HPP-L* with the main station *R*-J. The achieving by *HPP-L* a preset of 940 kW takes 1000 s but it can be concluded that the power flow of the observed line *L-62* changes the direction from load demand to generation after approximately 150 s when the generation of the *HPP-L* obtains a level of 500 kW. Switching on the battery energy storage *ESS-L* additionally increases the level of transmitted generation power by the observed line. Naturally, the observed process has a positive impact on decreasing the load demand of the transformer in the main station *R-J*. Figure 7 presents active power changes in the high voltage/medium voltage (HV/MV) transformer in the main station *R-J* during the switching operation series of *HPP-L*, *PV-C* and *ESS-L*.

With regards to the network requirements for the voltage changes caused by the integration of the DERs with the electrical power systems presented in Section 2, the assessment of the influence of the simulated series of the switching operations of *HPP-L*, *PV-C* and *ESS-L* on the voltage changes in the connection point of the DER, as well as on the secondary side of the HV/MV transformer, is presented. When observing Figures 8 and 9, it may be indicated that inserting the active power from *HPP*-L into the associated line *L-62* causes a slow voltage increase at the 20 kV busbar of the hydro power plant on the level of 1%. At the same time, the voltage on the 20 kV busbars at the main station *R-J* changes by less than 0.1%. When the *ESS-L* generates about 500 kW, there is an increase in voltage at 20 kV busbar of the connection point of *HPP-L* and *ESS-L* approximately on the level of 0.6% and a slight change in voltage. A sudden switching off of the *HPP-L* and *ESS-L* causes a rapid voltage change at the connection point of these energy sources on the level of 1.6% and simultaneously a rapid voltage change of less than 0.1% at the 20 kV busbar of the main station *R-J*.

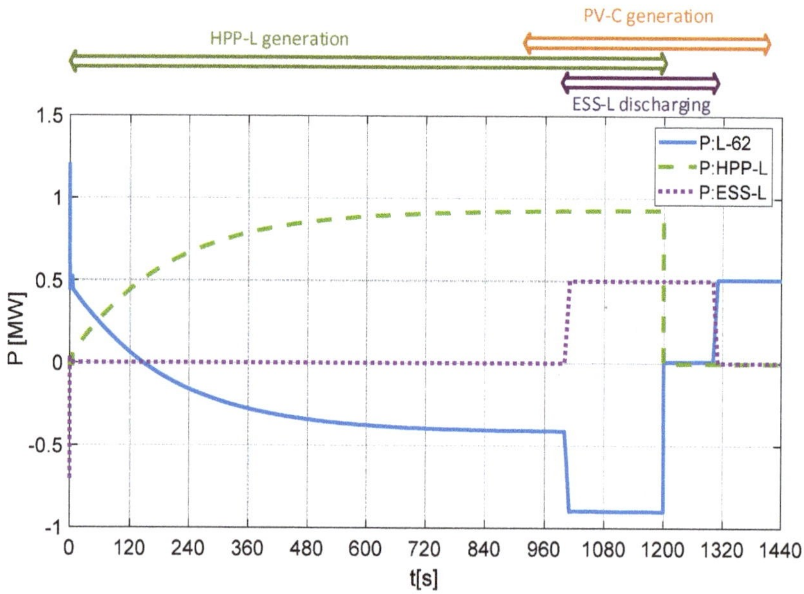

Figure 6. Active power changes *P* in the distribution line *L-62* associated with main station *R-J*. Changes of generated power of hydropower plant *HPP-L*. Changes of generated power of battery storage system *ESS-L*. Analysis carried out during a series of switching operations of *HPP-L*, *PV-C* and *ESS-L*.

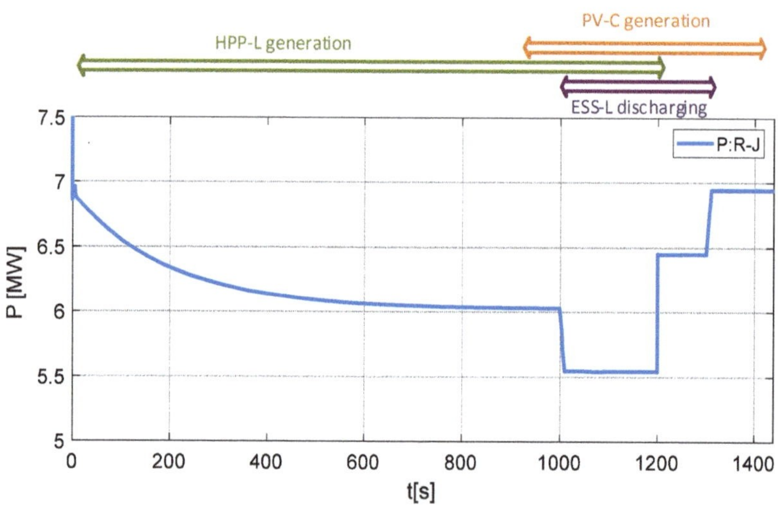

Figure 7. Active power changes *P* in the high voltage/medium voltage (HV/MV) transformer in the main station *R-J* during the series of switching operations of *HPP-L*, *PV-C* and *ESS-L*.

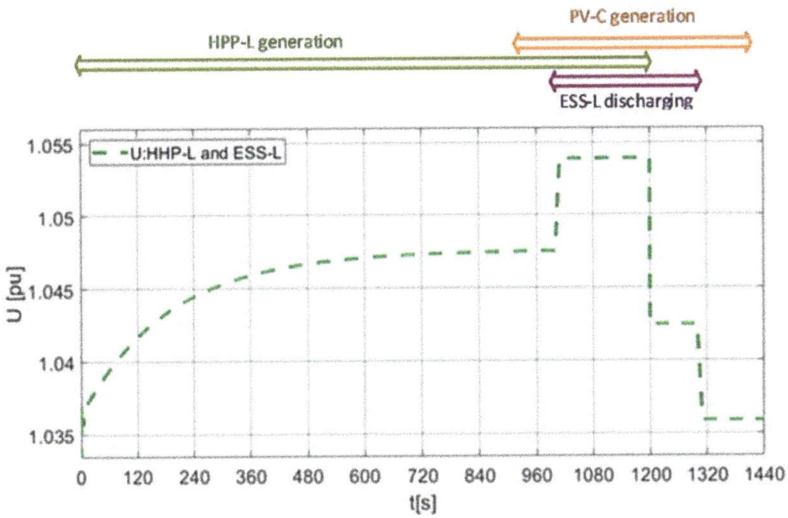

Figure 8. Voltage changes *U* in the connection point of the hydropower plant *HPP*-L and battery energy storage system *ESS-L* during the series of switching operations of *HPP-L, PV-C* and *ESS-L*.

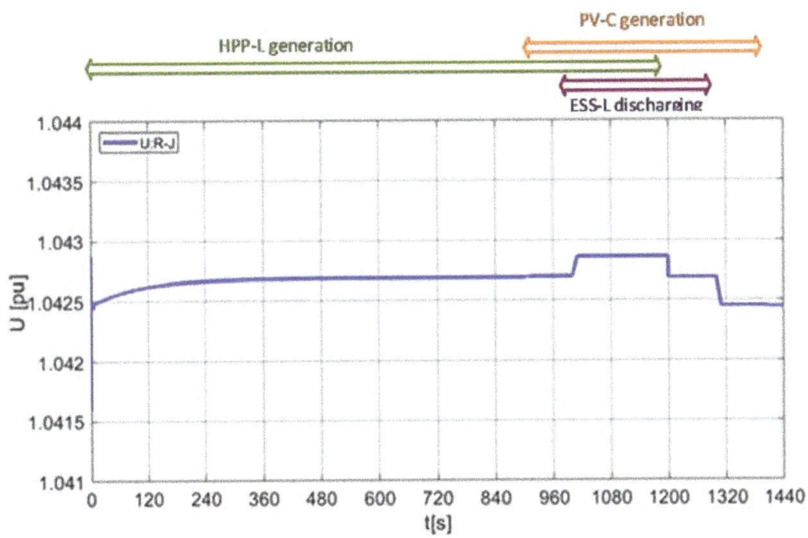

Figure 9. Voltage changes *U* at the secondary side of the HV/MV transformer in the main station *R-J* during the series of switching operations of *HPP-L, PV-*C and *ESS-L*.

When observing voltage changes at the secondary side of the HV/MV transformer in the main station *R-Z* presented in Figure 10, it may be concluded that the generation of active power in analyzed network connected to the main station *R-J* is slightly noticeable at the busbar of station *R-Z* which is associated with station *R-J* by high voltage line. Additionally, switching on the photovoltaic installation *PV-C* with 100 kW of active power causes a slight change in voltage at the 20 kV busbar of the main station *R-Z* that is not noticeable in the station *R-J*. However, it should be emphasized that the observed changes are small and at the level of one-hundredth of a percent of the 20 kV nominal value.

Noticed voltage changes can be compared with the quoted requirement of voltage changes on the level of three percentage and permissible value of voltage level not more than 10% of nominal value [60].

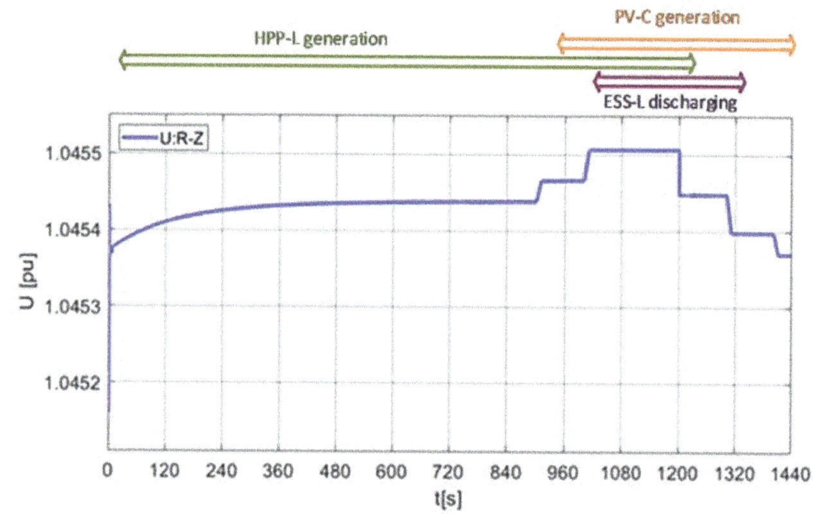

Figure 10. Voltage changes U at the secondary side of the HV/MV transformer in the main station *R-Z* during the series of switching operations of *HPP-L*, *PV-C* and *ESS-L*.

The simulations aimed to identify the direct impact of the active power changes generated by the HPP and BES on the load reduction in the distribution line, as well as on the voltage changes in the node of the connection and the substation. Therefore, standard methods of voltage regulations, including the on-load-tap-changer, were not used in the simulations. In reality, the tap-changer control is implemented in the main substations 110/20 kV, denoted as *R-J* and *R-Z*. However, the classical regulation strategy for the HV/MV transformer often uses a step of regulation on the level of 1.09–1.10%. In presented simulations indicated changes of voltage level in the main substations caused by BES or *PV-C* were less than the classical step of on-load-tap-changer regulation.

4.3. Identification of the Maximum Power Capacity of the ESS in the Considered Node of the VPP Regarding the Power Quality Voltage Profile

In terms of VPP efficiency and sensitivity, it is important to identify the maximum level of ESS power capacity that can be connected to the planned node. In order to identify the maximum power capacity of the ESS, it is proposed to conduct investigations with power quality parameters of the grid and requirements for the integration of the generation units with power systems. The impact of ESS power capacity on economic efficiency is considered in paper [36]. In this paper, the maximum power capacity of *ESS-L* is identified using a simplified analytic derivation, as well as Matlab modeling and simulation.

A rough estimation of the maximum power capacity of the considered battery energy storage *ESS-L* connected to the same node of MV network as hydro power plant *HPP*-L can be calculated based on short circuit power related to the connection point of *ESS-L* and *HPP-L*. The simplified one-phase Thevenin's equivalent circuit, which can be used for rough calculations, is presented in Figure 11.

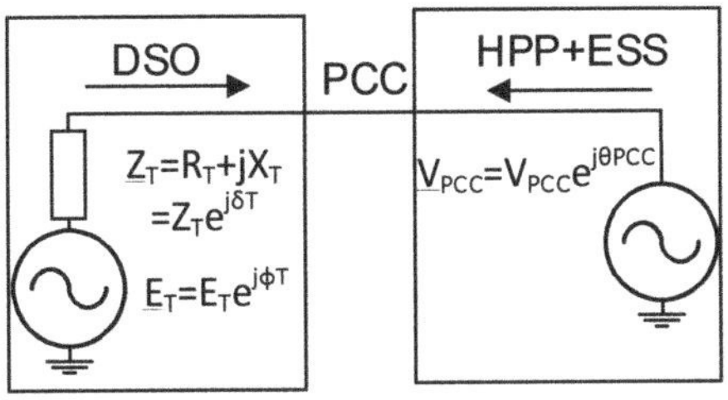

Figure 11. Simplified one-phase Thevenin's equivalent circuit used for a rough estimation of the influence of power capacity of considered hydro power plant and energy storage system on voltage level.

In a simplified estimation of the influence of the selected generation unit on the voltage condition in the connection point, a critical simplification can be considered. Firstly, the investigated network is treated as unloaded so that the decrease of voltage caused by the load is not taken into consideration. Only the direct influence of the considered generation is then revealed. As a result of the mentioned assumption, before the connection of the power unit, the Thevenin's substitute voltage source magnitude E_T in the point of common coupling (PCC) is the equal nominal voltage. The Thevenin's reactance X_T is equal to short circuits reactance X_Q addressed to the node of the connection point. The resistance of Thevenin's equivalent can be neglected in comparison to reactance. The parameters of Thevenin's equivalent circuits can be calculated as:

$$E_T = \frac{U_N}{\sqrt{3}} \tag{2}$$

$$X_T = X_Q = \frac{c \cdot U_N^2}{S_{kQ}}, \tag{3}$$

where:

- U_N—the nominal voltage phase to phase value,
- S_{kQ}—the short circuit apparent power in the connection node,
- C—the short circuit factor (c = 1 for minimal short circuit power, c = 1.1 for maximal short circuit power).

Due to the high influence of reactive power on the voltage level, the second critical assumption in the simplified calculation is that the *HPP-L* and *ESS-L* only generate a reactive power in the PCC. The voltage change is caused by a voltage associated with the short circuit reactance and current flow I_{PCC} inserted into the network by both generating units connected to the PCC operating at maximum power. The estimated steady state voltage change visible in the PCC can be expressed by:

$$\Delta U_C = \sqrt{3} \cdot I_{PCC} \cdot X_Q = \sqrt{3} \cdot \frac{S_{PCC}}{\sqrt{3} \cdot U_N} \cdot \frac{c \cdot U_N^2}{S_{kQ}} = c \cdot U_N \cdot \frac{S_{PCC}}{S_{kQ}}, \tag{4}$$

where: S_{PCC}—the maximum power capacity of the power generation unit connected to the PCC, which in the described case study is a sum of generated power HPP and ESS—$S_{PCC} = S_{HHP} + S_{ESS}$.

Combining definition of voltage change d_C introduced in Equation (1) with Equation (4) allows deriving a direct relation between maximum power capacity of considered power generation unit connected to the PCC (S_{PCC}) with the short circuit power which characterizes equivalent of the network visible in point of the PCC (S_{kQ}). This relation can be revealed as:

$$d_c = \frac{|\Delta U_C|}{U_N} = c \cdot \frac{S_{PCC}}{S_{kQ}}. \tag{5}$$

Equation (5) can be recalculated in order to express the maximum power capacity of the power generating unit connected to the considered PCC which is characterized by short circuit power. Short circuit power depends on the permissible level of rapid voltage change.

$$S_{PCC} = \frac{d_c}{c} \cdot S_{kQ}. \tag{6}$$

Short circuit power in the selected node of the investigated power network, that is, in the busbar of the main station *R-J* and in the connection point of the hydropower plant *HPP-L* and battery energy storage system *ESS-L*, are presented in Table 4. Next, taking into account permitted levels of rapid voltage change d_c = 3% and short circuit factor of c = 1.1 as quoted in Section 2, it is possible to use Equation (6) to estimate the maximum capacity of the power generating units that can be connected to the investigated node of the power network from the point of view of rapid voltage change requirement. An example of the calculation, in relation to the main power station *R-J* and the connection point of the hydropower plant and energy storage system (node L), is compared in Table 4. It can be concluded that the nodes located deep in the power grid are characterized by a lower level of short circuit power which ultimately increases the limitation of the capacity of the generating unit that can be connected in that node. When referring to the connection node of the hydropower plant and energy storage system, which is characterized by short circuit power on the level of 54.2 MVA, the maximum power capacity of the generation unit is limited to 1.48 MW. Assuming the operation of the hydropower plant *HPP-L* with a maximum power level of 0.94 MW, it can be concluded that the permissible power of the energy storage system *ESS-L* connected to the same node is limited to 0.54 MW. The presented calculation results of the possible power of the battery energy storage *ESS-L* should be treated as a rough estimation. The results are extremely limited by the simplification of the network, the unloaded condition, the reactive power consideration and the restricted limit of the rapid voltage change d_c = 3%.

Table 4. Short circuit power in the selected node of the investigated power network and the estimated maximum power capacity of the power generation unit permissible in terms of rapid voltage change requirements.

Node of the Investigated Power Network	Short Circuit Power S_{kQ} [MVA]	Estimated Maximum Power Capacity of Power Generation Unit S_{PCC} [MW]
Main station *R-J*	209.3 MVA	5.71
PCC of *HPP-L* and *ESS-L*	54.2 MVA	1.48

In order to obtain a more precise estimation of the desired value of the maximum power capacity of the *ESS-L*, a simulation of the influence of gradually increasing the power of the *ESS-L* on the voltage level in the connection point of the *HPP-L* and *ESS-L* is proposed. The basic conditions of the simulation are similar to those previously used when the effect of switching on the DER series on the voltage level was simulated. These preliminary assumptions are as follows—the initial power flows relate to summer peak load demand and the *HHP-L* power generation level is a maximum of around 940 kW. The result of the simulation is presented in Figure 12. The results allow concluding that from the point of view of acceptable rapid voltage changes at the point of connection of *HPP-L* and *ESS-L*, the total maximum capacity of these two generating units should be in the range from 2 MW to 2.4 MW. Assuming that the power generated by *HPP-L* is around 1 MW, it can be concluded that the possible maximum capacity of a given energy storage system *ESS-L* is limited to 1 MW or 1.4 MW. In comparison with the method based on simplified calculations using short circuit power circuits, the result obtained using more precise simulations is more realistic.

The rough estimation using short circuit power is the fast method, however, usually gives relatively underestimated results. The short-circuit equivalent model is dedicated to the linear electrical components. In addition, the short circuit current is significantly modified in the presence of power electronic inverters used to integrate DERs and ESSs into the power supply system. The results of the calculation of the maximum power of a given ESS using network modeling are more technically realistic but require modeling and computing power. It should be mentioned that the estimation of the maximum power capacity of a considered ESS based on modeling and simulation is more accurate as it includes:

- regulation of the transformer in main power stations *R-J* and *R-Z*,
- the power exchange between main power stations *R-J* and *R-Z* with a 110 kV line,
- parameters of the individual line sections connecting DER with main stations,
- parameters of the loads distributed along the lines,
- parameters of the DERs, especially the contribution of the power inverters in short circuits, possible dynamics during fast load changes, speed limits for power changes.

The presented results were used in the accompanying paper [36] in the point concerning the economic efficiency test where a 0.5 MW or 1 MW battery energy storage system is considered to be used in the VPP topology. In Reference [36] general aspects related to VPP concepts were also examined, including the analysis of the advantages and disadvantages of using one ESS compared to many small ESSs or more RES.

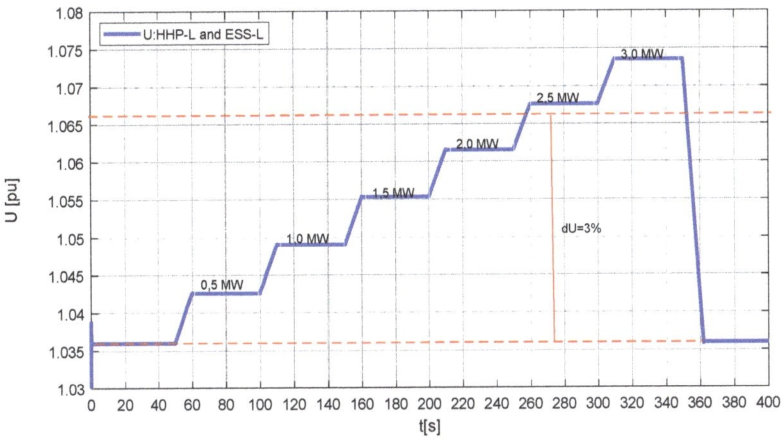

Figure 12. Voltage changes at the point of connection of MV *HPP-L* and *ESS-L* during the gradual increase of maximum power of power generators.

4.4. *Study on the Impact of the Power Control Strategy Used in a PV Power Plant on the Resources Available for VPP*

Section 3.2 discusses the relationship between active power and reactive power applied to low voltage generation units. It has been shown that such generating units have a standard *cosφ(P)* characteristic assuming a reactive power consumption for the production of active power above 50% of the maximum power of the generating unit. This regulation strategy serves to reduce of voltage increase. However, from the point of view of the virtual power plant, the regulation introduces certain restrictions which affect the availability of resources integrated into the VPP. In order to highlight this issue, studies of PV power plants belonging to the VPP were carried out. The PV power plant mentioned is a 132 kW PV installation marked as PV-C in Figure 4 and Table 2. This power plant is the facility consisting of several individual photovoltaic power plants PV, three-phase and single-phase installations using different installed power and different photovoltaic technologies but all connected in the same PCC. From a PCC point of view, the combined phase is occupied almost symmetrically but there may be some differences between the phases. The aim of the presented studies is to determine the level of energy, which is redirected to the reactive power consumption instead of the active power generation. In order to achieve this, the actual measurement of changes in active and reactive power in the selected week, which is characterized by similar daily weather conditions, was analyzed. Changes in active and reactive power during the test week due to changes in solar irradiance are shown in Figure 13. It can be seen that a high level of active power generation is accompanied by reactive power consumption, which indicates that PV installations integrated with the PV power plant realize the *cosφ(P)* characteristic. In order to emphasize the observed correlation, Figure 14 shows the correlation between solar irradiation and active and reactive power. The calculated Pearson correlation coefficients for active and reactive power confirm the high correlation between solar irradiation and the generation of active power and, consequently, reactive power consumption.

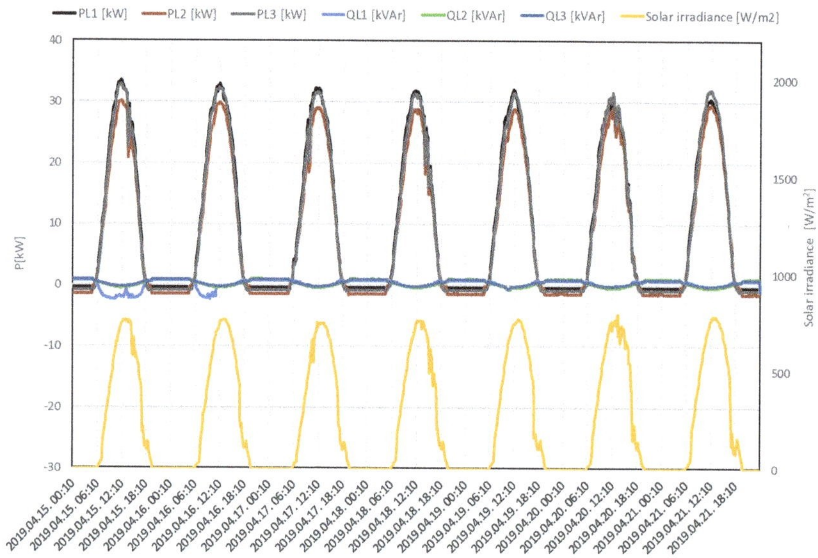

Figure 13. Changes in active (P_{L1}, P_{L2}, P_{L3}) and reactive (Q_{L1}, Q_{L2}, Q_{L3}) power in specific phases in point of common coupling (PCC) of PV power plant PV-C and changes in solar irradiation in the examined week.

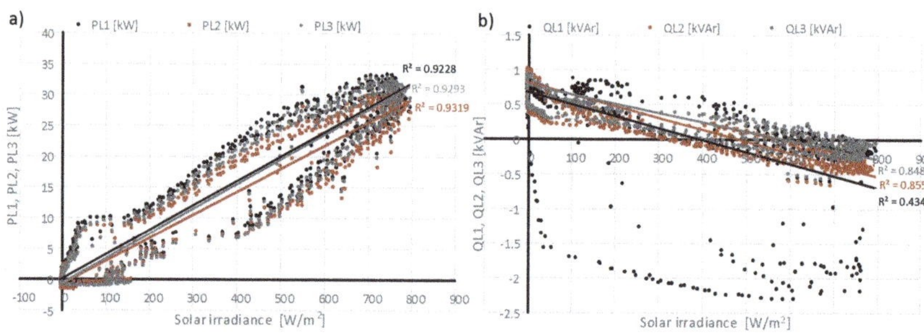

Figure 14. Correlation analysis between solar irradiation and active power (**a**) and reactive power (**b**) in PCC of PV power plant PV-C during the investigated week.

Table 5 contains an analysis of active power generation and reactive power consumption over one week to determine the effect of the implemented *cosφ(P)* characteristic on the reduction of active power generation availability. It was shown that during one observation week, which is characterized by sunny weather, the total amount of active energy produced is 4799.91 [kWh] and at the same time, reactive energy consumed is 53.87 [kvarh]. This leads to the conclusion that during the sunny week

1.11% of the total energy from the considered PV is not available for the VPP planning because it is used for reactive power regulation. The result presented corresponds to a sunny week, therefore a long-term analysis should be carried out to verify this observation.

Table 5. Analysis of energy production of the PV power plant during the investigated week in order to determine the impact of reactive power regulation on reducing the availability of the resources integrated into the VPP in the perspective of the observed week.

Parameter	Phase L1	Phase L2	Phase L3	Total
Time of active power generation [h]/[% 168 h of the week]	92.17 [h] 54.86 [%]	93.17 [h] 55.46% [%]	87.50 [h] 52.08 [%]	90.95 [h] (mean) 54.13 [%] (mean)
Time of reactive power consumption [h]/[% 168 h of the week]	52.33 [h] 31.15 [%]	51.50 [h] 30.65%	38.00 [h] 22.62 [%]	47.28 (mean) 28.14 [%] (mean)
Generated active energy [kWh]/[% of total energy]	1666.51	1509.35	1624.05	4799.91 [kWh] 98.89 [%]
Consumed reactive energy [kvarh]/[% of total energy]	34.18	14.14	5.55	53.87 [kvarh] 1.11 [%]

5. Discussion

This paper formulates a thesis that requirements related to regulation, protection and integration of power generation units and battery storage system with power systems have an impact on the planning and operation strategies of the VPP. After a comprehensive study of the current grid codes, standards and papers, three topics were selected and highlighted in the VPP:

- identification of an operational condition of the 1 MW hydro power plant and 0.5 MW battery energy storage connected to the same node and its impact on power flow and voltage level in observed network covered by VPP,
- identification of maximum power capacity of battery energy storage, which can be connected to a given node and identification of the overall capacity of the examined fragment of the distribution network to connect possible DERs/ESSs
- identification of the impact of power control strategy applied to a PV power plant on resources available for the VPP.

Mentioned issues were investigated based on the real VPP project. The results obtained are distinctive and related to the investigated localization of the VPP, however, the proposed test method and results obtained can be treated more generally. In order to determine indicated aspects, a few test methods using Matlab simulation have been implemented, combining actual load demand measurements with data generated as a prerequisite for simulation and direct analysis of actual measurements. Application of the proposed method to the real case study of the VPP allowed to formulate several observations related to the operational capabilities of the VPP and technical requirements related to the integration of the DER and ESS with the utility grid:

- Grid capacity for connection of the DERs and ESSs is limited. One of the main limitations is the requirement for rapid voltage changes. In the presented studies it was shown that using modeling of the considered network covered by the planned VPP, it is possible to determine the maximum power capacity of the ESS intended to be connected in the selected node. It has also been proven that the use of the simplified model adopted for short-circuit calculation is extremely simplified and results of rough estimates usually return relatively underestimated results.
- The implementation of *cosφ(P)* characteristic for reactive power control in power inverters integrating PVs with the grid reduces the availability of active power generation, which ultimately means restrictions on the use of PV in the VPP planning strategy. The presented results of real PV measurements associated with the considered VPP showed that 1.11% of the total energy from the considered PV is not available for VPP planning in the sunny week because it is designed to regulate reactive power.

- Representation of the grid covered by the VPP using a simulation model, which is complemented by actual measurements, provides extended opportunities for research into the choice of strategy for planning and operating a VPP. Presented results allowed to determine the impact of power changes of energy storage system or hydropower plant on the reduction of load on lines and transformers and on voltage changes. It can be used to create energy system services provided by VPP.

Obtained results constitute the influence of the technical requirements for DER and ESS integration with the power grid on the VPP operational condition.

6. Conclusions

Technical aspects related to the integration of DER with the power systems can be treated as boundary conditions for the VPP planning and operation strategies. This article presents indicated results obtained for a specific case of a VPP and a broad generalization for another VPP location is not easy to achieve. However, the presented assessment method, the methodology of the studies and investigations, can be adapted and applied to another VPP topology. In addition, the definition of limitations on VPP resources formulated by the technical aspects can be further used as a prerequisite for economic research. For example, based on the studies carried out, it was indicated that planned battery energy storage in the investigated grid covered by the VPP could be increased from the planned capacity of 0.5 MW to 1.0 MW. Therefore, this result allowed us to investigate the impact of the size of the energy storage system to economic efficiency in associated work [36].

Author Contributions: Conceptualization, T.S., E.R.-S., M.W.; methodology of technical aspects, T.S., M.S., W.R.; methodology of economic aspects, E.R.-S., M.W.; software for technical simulation, M.S.,W.R., J.S.; software for economic calculation, E.R.-S.; formal analysis in content of technical aspects, T.S., M.J., R.L., J.R., D.K., P.K.; formal analysis in content of economic aspects, E.R.-S., M.W.; investigation in the range of technical aspects, T.S., W.R., M.S., M.J., B.S.; investigation in the range of economic aspects, E.R.-S., M.W.; resources of technical aspects, D.K., R.L., J.R., M.J., T.S., Z.L., P.K.; resources of economic aspects, E.R.-.S., M.W.; technical data curation, J.S., M.S., D.B., P.J.; economic data curation, E.R.-S., M.W.; visualization of technical aspects, W.R., M.J., D.K., P.K.; visualization of economic aspects E.R.-S., M.W.; writing—original draft preparation, T.S., M.J., D.K., J.R., R.L. and E.R.-S., M.W.; writing—review and editing, T.S., M.J., E.R.-S., M.W.; supervision, T.S., M.W.; project administration, T.S., P.J.; funding acquisition, T.S., P.J. All authors have read and agreed to the publisher version of the manuscript.

Funding: This research was funded by the National Center of Research and Development in Poland, the project "Developing a platform for aggregating generation and regulatory potential of dispersed renewable energy sources, power retention devices and selected categories of controllable load" supported by European Union Operational Programme Smart Growth 2014-2020, Priority Axis I: Supporting R&D carried out by enterprises, Measure 1.2: Sectoral R&D Programmes, POIR.01.02.00-00-0221/16, performed by TAURON Ekoenergia Ltd.

Acknowledgments: The research uses annual weather conditions and power production data of PV power plant provided by Center of Energy Technologies in Świdnica, Poland.

Conflicts of Interest: The authors declare no conflict of interest.

Abbreviations

The following abbreviations are used in the paper:

ACER	Agency for the Cooperation of Energy Regulators
BMS	battery management system
c	short circuit factor
CHP	combined heat and power
CIGRE	Conseil International des Grands Réseaux Électriques
CIM	common information model
DC	direct current
DER, DG	distributed energy resources, distributed generation
DSO	distribution system operator
ΔU_C	steady state voltage change
d_C	relative maximum steady state voltage change
$\cos\phi$	power factor

ENTSO-E	European Network of Transmission System Operators for Electricity
EPS	electrical power system
ESS	energy storage system
f	frequency
g	gravitational acceleration
H	height
HPP	hydro power plant
HV	high voltage
ICT	information and communication technology
I_{PCC}	current inserted in the point of common coupling by generation unit
LV	low voltage
MV	medium voltage
NC	network code, grid code
ode24	function in Matlab for solvation a ordinary differential Equation
P	active power
P_{max}	maximum power capacity
P_M	actual power
PCC	point of common coupling
PQ	power quality
Q	reactive power
RES	renewable energy sources
S_{kQ}	short circuit apparent power
SoC	state of charge
S_{PCC}	power of the generation unit connected to the point of common coupling
ST1A	IEEE type ST1A excitation system mode in Matlab
U	voltage
U_N	nominal voltage
U_C	steady state voltage
VPP	virtual power plant
X_Q	short circuit reactance
X_T	reactance of Thevenin's circuit equivalent

References

1. Awerbuch, S.; Preston, A. *The Virtual Utility*; Awerbuch, S., Preston, A., Eds.; Springer: Boston, MA, USA, 1997; Volume 26, ISBN 978-1-4613-7827-3.
2. Lis, R.; Czechowski, R. Transformation of microgrid to virtual power plant. In *Variability, Scalability and Stability of Microgrids*; Institution of Engineering and Technology: London, UK, 2019; pp. 99–142.
3. Othman, M.M.; Hegazy, Y.; Abdelaziz, A.Y. A review of virtual power plant definitions, components, framework and optimization. *Int. Electr. Eng. J.* **2015**, *6*, 2010–2024.
4. Ghavidel, S.; Li, L.; Aghaei, J.; Yu, T.; Zhu, J. A review on the virtual power plant: Components and operation systems. In Proceedings of the 2016 IEEE International Conference on Power System Technology (POWERCON), Wollongong, Australia, 28 September–1 October 2016; pp. 1–6.
5. Richter, A.; Hauer, I.; Wolter, M. Algorithms for technical integration of virtual power plants into German system operation. *Adv. Sci. Technol. Eng. Syst. J.* **2018**, *3*, 135–147. [CrossRef]
6. Yavuz, L.; Önen, A.; Muyeen, S.M.; Kamwa, I. Transformation of microgrid to virtual power plant—A comprehensive review. *IET Gener. Transm. Distrib.* **2019**, *13*, 2077–2087. [CrossRef]
7. Etherden, N.; Vyatkin, V.; Bollen, M.H.J. Virtual power plant for grid services using IEC 61850. *IEEE Trans. Ind. Inform.* **2016**, *12*, 437–447. [CrossRef]
8. Asmus, P. Microgrids, virtual power plants and our distributed energy future. *Electr. J.* **2010**. [CrossRef]
9. Vergados, D.J.; Mamounakis, I.; Makris, P.; Varvarigos, E. Prosumer clustering into virtual microgrids for cost reduction in renewable energy trading markets. *Sustain. Energy Grids Netw.* **2016**. [CrossRef]
10. Anoh, K.; Maharjan, S.; Ikpehai, A.; Zhang, Y.; Adebisi, B. Energy peer-to-peer trading in virtual microgrids in smart grids: A game-theoretic approach. *IEEE Trans. Smart Grid* **2020**. [CrossRef]

11. Adu-Kankam, K.O.; Camarinha-Matos, L.M. Towards collaborative virtual power plants: Trends and convergence. *Sustain. Energy Grids Netw.* **2018**, *16*, 217–230. [CrossRef]
12. Alahyari, A.; Ehsan, M.; Mousavizadeh, M.S. A hybrid storage-wind virtual power plant (VPP) participation in the electricity markets: A self-scheduling optimization considering price, renewable generation, and electric vehicles uncertainties. *J. Energy Storage* **2019**, 25. [CrossRef]
13. Thavlov, A.; Bindner, H.W. Utilization of flexible demand in a virtual power plant set-up. *IEEE Trans. Smart Grid* **2015**, *6*, 640–647. [CrossRef]
14. Kardakos, E.G.; Simoglou, C.K.; Bakirtzis, A.G. Optimal offering strategy of a virtual power plant: A stochastic bi-level approach. *IEEE Trans. Smart Grid* **2016**, *7*, 794–806. [CrossRef]
15. Tang, W.; Yang, H.T. Optimal operation and bidding strategy of a virtual power plant integrated with energy storage systems and elasticity demand response. *IEEE Access* **2019**, *7*, 79798–79809. [CrossRef]
16. Papadaskalopoulos, D.; Pudjianto, D.; Strbac, G. Decentralized coordination of microgrids with flexible demand and energy storage. *IEEE Trans. Sustain. Energy* **2014**, *5*, 1406–1414. [CrossRef]
17. Sadeghian, O.; Shotorbani, A.M.; Mohammadi-Ivatloo, B. Generation maintenance scheduling in virtual power plants. *IET Gener. Transm. Distrib.* **2019**, *13*, 2584–2596. [CrossRef]
18. Han, N.; Wang, X.; Chen, S.; Cheng, L.; Liu, H.; Liu, Z.; Mao, Y. Optimal configuration of energy storage systems in virtual power plants including large-scale distributed wind power. *IOP Conf. Ser. Earth Environ. Sci.* **2019**, *295*, 042072. [CrossRef]
19. Abdolrasol, M.G.M.; Hannan, M.A.; Mohamed, A.; Amiruldin, U.A.U.; Abidin, I.B.Z.; Uddin, M.N. An optimal scheduling controller for virtual power plant and microgrid integration using the binary backtracking search algorithm. *IEEE Trans. Ind. Appl.* **2018**, *54*, 2834–2844. [CrossRef]
20. Liang, Z.; Alsafasfeh, Q.; Jin, T.; Pourbabak, H.; Su, W. Risk-constrained optimal energy management for virtual power plants considering correlated demand response. *IEEE Trans. Smart Grid* **2019**, *10*, 1577–1587. [CrossRef]
21. Kasaei, M.J.; Gandomkar, M.; Nikoukar, J. Optimal management of renewable energy sources by virtual power plant. *Renew. Energy* **2017**, *114*, 1180–1188. [CrossRef]
22. Hannan, M.A.; Abdolrasol, M.G.M.; Faisal, M.; Ker, P.J.; Begum, R.A.; Hussain, A. Binary particle swarm optimization for scheduling mg integrated virtual power plant towards energy saving. *IEEE Access* **2019**, *7*, 1. [CrossRef]
23. Koraki, D.; Strunz, K. Wind and solar power integration in electricity markets and distribution networks through service-centric virtual power plants. *IEEE Trans. Power Syst.* **2018**, *33*, 473–485. [CrossRef]
24. Bagchi, A.; Goel, L.; Wang, P. Adequacy assessment of generating systems incorporating storage integrated virtual power plants. *IEEE Trans. Smart Grid* **2019**, *10*, 3440–3451. [CrossRef]
25. Chen, S.; Zhang, T.; Gooi, H.B.; Masiello, R.D.; Katzenstein, W. Penetration rate and effectiveness studies of aggregated BESS for frequency regulation. *IEEE Trans. Smart Grid* **2016**, *7*, 167–177. [CrossRef]
26. Zamani, A.G.; Zakariazadeh, A.; Jadid, S.; Kazemi, A. Stochastic operational scheduling of distributed energy resources in a large scale virtual power plant. *Int. J. Electr. Power Energy Syst.* **2016**, *82*, 608–620. [CrossRef]
27. Nosratabadi, S.M.; Hooshmand, R.A.; Gholipour, E. A comprehensive review on microgrid and virtual power plant concepts employed for distributed energy resources scheduling in power systems. *Renew. Sustain. Energy Rev.* **2017**, *67*, 341–363. [CrossRef]
28. Yu, S.; Fang, F.; Liu, Y.; Liu, J. Uncertainties of virtual power plant: Problems and countermeasures. *Appl. Energy* **2019**, *239*, 454–470. [CrossRef]
29. Survey, P.S.A.; Rakhshani, E.; Rouzbehi, K.; Adolfo, J.S.; Tobar, A.C. Integration of large scale PV-based generation into power systems: A survey. *Energies* **2019**, *12*, 1425. [CrossRef]
30. Ribeiro, C.; Pinto, T.; Vale, Z.; Baptista, J. Remuneration and tariffs in the context of virtual power players. In *Trends in Cyber-Physical Multi-Agent Systems, Proceedings of the The PAAMS Collection-15th International Conference, PAAMS, Porto, Portugal, 21–23 June 2017*; la Prieta, F., Vale, Z., Antunes, L., Pinto, T., Campbell, A.T., Julián, V., Neves, A.J.R., Moreno, M.N., Eds.; Springer International Publishing: Cham, Switzerland, 2018; pp. 284–286. ISBN 978-3-319-61578-3.
31. Boldt, D.; Faria, P.; Vale, Z. Integration of pumping in virtual power players management considering demand response. In Proceedings of the 13th International Conference on the European Energy Market (EEM), Porto, Portugal, 6–9 June 2016; pp. 1–5.

32. Silva, C.; Faria, P.; Vale, Z. Multi-period observation clustering for tariff definition in a weekly basis remuneration of demand response. *Energies* **2019**, *12*, 1248. [CrossRef]
33. Faria, P.; Spínola, J.; Vale, Z. Distributed energy resources scheduling and aggregation in the context of demand response programs. *Energies* **2018**, *11*, 1987. [CrossRef]
34. Ribeiro, C.; Pinto, T.; Vale, Z.; Baptista, J. Data mining for prosumers aggregation considering the self-generation. In *Distributed Computing and Artificial Intelligence, Proceedings of the 14th International Conference, Porto, Portugal, 21–23 June 2017*; Omatu, S., Rodríguez, S., Villarrubia, G., Faria, P., Sitek, P., Prieto, J., Eds.; Springer International Publishing: Cham, Switzerland, 2018; pp. 96–103. ISBN 978-3-319-62410-5.
35. Faria, P.; Vale, Z. A demand response approach to scheduling constrained load shifting. *Energies* **2019**, *12*, 1752. [CrossRef]
36. Sikorski, T.; Jasiński, M.; Ropuszyńska-Surma, E.; Węglarz, M.; Kaczorowska, D.; Kostyła, P.; Leonowicz, Z.; Lis, R.; Rezmer, J.; Rojewski, W.; et al. A case study on distributed energy resources and energy-storage systems in a virtual power plant concept: Economic aspects. *Energies* **2019**, *12*, 4447. [CrossRef]
37. Yang, J.; Huang, Y.; Wang, H.; Ji, Y.; Li, J.; Gao, C. A regulation strategy for virtual power plant. In Proceedings of the 2017 4th International Conference on Systems and Informatics (ICSAI), Hangzhou, China, 11–13 November 2017; pp. 375–379. [CrossRef]
38. Roossien, B.; Hommelberg, M.; Warmer, C.; Turkstra, J.-W. Virtual power plant field experiment using 10 micro-CHP units at consumer premises. *IET Conf. Proc.* **2008**, 5. [CrossRef]
39. Unger, D.; Spitalny, L.; Myrzik, J.M.A. Voltage control by small hydro power plants integrated into a virtual power plant. In Proceedings of the 2012 IEEE Energytech, Cleveland, OH, USA, 29–31 May 2012; pp. 1–6. [CrossRef]
40. Beguin, A.; Nicolet, C.; Kawkabani, B.; Avellan, F. Virtual power plant with pumped storage power plant for renewable energy integration. In Proceedings of the 2014 International Conference on Electrical Machines (ICEM), Berlin, Germany, 2–5 September 2014; pp. 1736–1742. [CrossRef]
41. Bilbao, J.; Bravo, E.; Rebollar, C.; Varela, C.; Garcia, O. Virtual power plants and virtual inertia. In *Power Systems*; Springer Verlag: Bilbao, Spain, 2020; pp. 87–113. ISBN 16121287.
42. Othman, M.M.; Hegazy, Y.G.; Abdelaziz, A.Y. Electrical energy management in unbalanced distribution networks using virtual power plant concept. *Electr. Power Syst. Res.* **2017**, *145*, 157–165. [CrossRef]
43. Nemati, M.; Zöller, T.; Tenbohlen, S.; Tao, L.; Mueller, H.; Braun, M. Optimal energy management system for future microgrids with tight operating constraints. In Proceedings of the 2015 12th International Conference on the European Energy Market (EEM), Lisbon, Portugal, 19–22 May 2015; pp. 1–5. [CrossRef]
44. Ali, J.; Massucco, S.; Silvestro, F.; Vinci, A. Participation of customers to virtual power plants for reactive power provision. In Proceedings of the 2018 53rd International Universities Power Engineering Conference (UPEC), Glasgow, UK, 4–7 September 2018; pp. 1–6. [CrossRef]
45. Dos Santos, T.V., Jr.; Bonatto, B.D.; Ferreira, C.; De Souza, Z.A.C. Impact of harmonic distortion on the energization of energy distribution transformers integrated in virtual power plants. In Proceedings of the 2018 15th International Conference on Control, Automation, Robotics and Vision, ICARCV, Singapore, 18–21 November 2018; pp. 1252–1256.
46. Braun, M. Virtual power plants in real applications-pilot demonstrations in Spain and England as part of the European project FENIX. In Proceedings of the ETG-Fachbericht-Internationaler ETG-Kongress, Düsseldorf, Germany, 27–28 October 2009.
47. Hess, T.; Schegner, P. Power schedule planing and operation algorithm of the local virtual power plant based on uCHP-devices. In Proceedings of the 2015 IEEE Power & Energy Society General Meeting, Denver, CO, USA, 26–30 July 2015; pp. 1–5.
48. Vuthi, P.P.; Lorenzen, P.; Schaefers, H.; Raths, S.; Krengel, S.; Sudeikat, J.; Thomsen, M. Smart power Hamburg: A virtual power plant for Hamburg. In Proceedings of the International ETG Congress-Die Energiewende-Blueprints for the New Energy Age, Bonn, Germany, 17–18 November 2015; pp. 1–8.
49. Binding, C.; Gantenbein, D.; Jansen, B.; Sundstrom, O.; Andersen, P.B.; Marra, F.; Poulsen, B.; Traeholt, C. Electric vehicle fleet integration in the danish EDISON project-a virtual power plant on the island of Bornholm. In Proceedings of the IEEE PES General Meeting, Providence, RI, USA, 25–29 July 2010; pp. 1–8.
50. Stock, A.; Bourne, G.; Brailsford, L.; Stock, P. *Fully Charged: Renewables and Storage Powering Australia*; Climate Council of Australia Limited: Potts Point, Australia, 2018.

51. CIGRE Conseil International des Grands Réseaux Électriques. *Connection Criteria at the Distribution Network for Distributed Generation*; CIGRE: Paris, France, 2007.
52. VDE Verband der Elektrotechnik Elektronik. *Power Generation Systems Connected to the Low-Voltage Distribution network: Technical Minimum Requirements for the Connection to and Parallel Operation with Low-Voltage Distribution Networks*; VDE-AR-N 4105:2011-08; VDE: Frankfurt, Germany, 2011.
53. Commission Regulation. *2016/631 of 14 April 2016 Establishing a Network Code on Requirements for Grid Connection of Generators*; The European Commission: Brussels, Belgium, 2016.
54. ACER Agency for the Cooperation of Energy Regulators. *Framework Guidelines On Electricity Grid Connections*; ACER: Ljubljana, Slovenia, 2011.
55. ENTSO-E European Network of Transmission System Operators for Electricity. *Requirements for Grid Connection Applicable to All Generators*; ENTSO-E: Brussels, Belgium, 2013.
56. Stadler, I. Study About International Standards for the Connection of Small Distributed Generators to the Power Grid. 2011. Available online: https://energypedia.info/images/temp/c/cf/20140508124849!phpvxPnex.pdf (accessed on 15 February 2020).
57. Sikorski, T.; Rezmer, J. Distributed generation and its impact on power quality in low-voltage distribution networks. In *Power Quality Issues in Distributed Generation*; InTech: London, UK, 2015.
58. Jasiński, M.; Rezmer, J.; Sikorski, T.; Szymańda, J. Integration monitoring of on-grid photovoltaic system: Case study. *Period. Polytech. Electr. Eng. Comput. Sci.* **2019**, *63*, 99–105. [CrossRef]
59. Klajn, A.; Bątkiewicz-Pantua, M. Application Note-Standard EN 50 160: Voltage Characteristics of Electricity Supplied by Public Electricity Networks. 2017. Available online: https://copperalliance.org.uk/uploads/2018/03/542-standard-en-50160-voltage-characteristics-in.pdf (accessed on 15 February 2020).
60. CENELEC Comité Européen de Normalisation Electrotechnique. *Voltage Characteristics of Electricity Supplied by Public Electricity Networks*; CENELEC: Brussels, Belgium, 2010.
61. Jasiński, M.; Sikorski, T.; Kostyła, P.; Kaczorowska, D.; Leonowicz, Z.; Rezmer, J.; Szymańda, J.; Janik, P.; Bejmert, D.; Rybiański, M.; et al. Influence of measurement aggregation algorithms on power quality assessment and correlation analysis in electrical power network with pv power plant. *Energies* **2019**, *12*, 3547. [CrossRef]
62. Zarębski, T. Analysis of the efficiency of energy storage systems. In Proceedings of the 2018 Innovative Materials and Technologies in Electrical Engineering (i-MITEL), Sulecin, Poland, 18–20 April 2018.
63. Waśkowicz, B.; Bojarski, J. Energy storage control algorithm for suppression of fluctuation of wind farm output power. In Proceedings of the 2018 Innovative Materials and Technologies in Electrical Engineering (i-MITEL), Sulecin, Poland, 18–20 April 2018; pp. 1–6.
64. Gubański, A.; Kaczorowska, D. Power flow optimization between microgrid and distribution system. In Proceedings of the 2018 Innovative Materials and Technologies in Electrical Engineering, i-MITEL, Sulecin, Poland, 18–20 April 2018; pp. 1–4.
65. Waltrich, G. Energy Management of Fast-Charger Systems for Electric Vehicles: Experimental Investigation of Power Flow Steering Using Bidirectional Three-Phase Three-Port Converters. Ph.D. Thesis, Technische Universiteit, Eindhoven, The Netherlands, 2013.
66. MIT Electric Vehicle Team A Guide to Understanding Battery Specification. Available online: http://web.mit.edu/evt/summary_battery_specifications.pdf (accessed on 15 February 2020).
67. Waclawek, Z.; Rezmer, J.; Janik, P.; Nanewortor, X. Sizing of photovoltaic power and storage system for optimized hosting capacity. In Proceedings of the 2016 IEEE 16th International Conference on Environment and Electrical Engineering (EEEIC), Florence, Italy, 7–10 June 2016. [CrossRef]
68. Kaczorowska, D.; Rezmer, J. Particle swarm algorithm for microgrid optimization. In Proceedings of the 2018 Innovative Materials and Technologies in Electrical Engineering, i-MITEL 2018, Tenerife, Spain, 10–12 April 2019; pp. 1–4.
69. Zhou, B.; Liu, X.; Cao, Y.; Li, C.; Chung, C.Y.; Chan, K.W. Optimal scheduling of virtual power plant with battery degradation cost. *IET Gener. Transm. Distrib.* **2016**, *10*, 712–725. [CrossRef]
70. PSE Polish Transmission System Operator Load of Polish Power System. Available online: https://www.pse.pl/web/pse-eng (accessed on 15 February 2020).

MDPI
St. Alban-Anlage 66
4052 Basel
Switzerland
Tel. +41 61 683 77 34
Fax +41 61 302 89 18
www.mdpi.com

Energies Editorial Office
E-mail: energies@mdpi.com
www.mdpi.com/journal/energies

www.ingramcontent.com/pod-product-compliance
Lightning Source LLC
Chambersburg PA
CBHW062104220926
51850CB00010B/2881

* 9 7 8 3 0 3 9 3 6 8 2 0 4 *